AF690855

PROTEAS
OF THE WORLD

PLATE 1: *Protea magnifica*, cream coloured variants of the Queen Protea

PROTEAS
OF THE WORLD

Lewis J. Matthews
Paintings by
Zoë Carter

Timber
Press

ACKNOWLEDGEMENTS

During the preparation of this book, and indeed over the years of the author's involvement with the Proteaceae, many people have assisted with information, advice and encouragement. It is impossible to list everyone, but the following are some to whom particular acknowledgement is due:

Kaye Matthews, the author's wife, for her encouragement and the hours spent checking manuscripts.
Dr John Rourke, Cape Town, South Africa, for his ongoing encouragement.
Mrs Ruth Middelmann, Cape Town, South Africa, for historical information.
Ian and Jocelyn Bell, Wanganui, New Zealand, for information.
Geoff Jewell, Te Horo, New Zealand, for information and encouragement.
Jack Clark, Auckland, New Zealand, for information.
Dr Philip Parvin, University of Hawaii at Manoa, Kula, Hawaii.

First published in North America in 1993 by Timber Press Inc., 9999 S.W. Wilshire, Suite 124, Portland, Oregon 97225, U.S.A.

in association with David Bateman Ltd,
32-34 View Road, Glenfield, Auckland, New Zealand

©1993 Lewis Matthews (text)
©1993 Zöe Carter (art)
©1993 David Bateman Ltd

All rights reserved. No part of this book may be reproduced in any form, by photostat, microfilm, or other means, or incorporated into any information retrieval system, electronic or mechanical, without the written permission of the copyright owner.

ISBN 0-88192-235-8

Design by Chris O'Brien
Typeset by Typeset Graphics Ltd
Printed in Hong Kong by
Cover illustration: *Protea* cv. 'Clark's Red' (lower flowers and bud), *Protea repens* (centre left), and *Protea aurea* (top).

PREFACE

Since the publication in 1983 of my first book on the Proteaceae *South African Proteaceae in New Zealand*, I have nursed an ambition to write another which would encompass the many magnificent genera from other regions, most particularly Australia.

While there have been and will be many other books written on the Proteaceae, some more botanically comprehensive, at the time of writing there has been no attempt to cover in one volume the most outstanding members of the family from various countries. Certainly, few have yet been published which use botanical paintings of such high calibre. It would be an almost impossible task to cover all the species and varieties of Proteaceae in one volume, especially using paintings; and even if it were achieved the cost of publication would probably be prohibitive. This book therefore seeks only to describe and illustrate a representative range of some of the most outstanding and more easily grown and obtainable genera, species and varieties. It is hoped that by so doing it will increase the gardening public's awareness and appreciation of this uniquely beautiful plant family.

Zoë Carter, whose wonderful paintings so enhanced *South African Proteaceae in New Zealand*, has, to my amazement, become even more adept with her paintbrushes, and her detailing of such complex flowers as banksias is quite remarkable. Her ability has inspired me to persevere with this book and to overcome some of the difficulties involved – notable a lack of time. Running a nursery which specialises in the Proteaceae takes up most of the hours in a day, and as our operation is virtually a two-person one in the form of a husband and wife team, squeezing in those extra hours to write has not been easy.

The family Proteaceae has been a source of wonderment for me since childhood when the striking differences in size and habit between the diminutive *Protea nana* and the almost gross King Protea, *P. cynaroides*, held me spellbound. Ever since that time in my parents' garden I have continued to be intrigued by this remarkable family of plants. For ten years or so, through my teenage years, the interest continued, and twenty years ago, after settling down to married life, the interest became a full-time business.

A significant change in our lifestyle occurred in 1985 when we relocated our nursery and home from Manakau in New Zealand's Horowhenua district of the North Island to French Farm Bay on the South Island's Banks Peninsula. The move has had numerous consequences for us, some planned for, some not, but few have been more far-reaching than the experiences of dealing with gardeners from areas receiving snow and heavier frosts than we were accustomed to. Gaining further knowledge at first hand from enthusiasts as to the frost-hardiness of various species and varieties in the garden has been particularly enlightening. Even our own property, although close to the sea of the Akaroa Harbour and having sloping characteristics to facilitate frost drainage, is still subject to occasional frosts of minus four degrees Celsius and several winters have seen a light snowfall. In spite of this we can grow almost all proteaceous plants in open situations without significant losses. Certain species, particularly *Protea magnifica* and *P. nana*, thrive far more than at our previous nursery.

It is my sincere wish that this book will encourage gardeners, if they have not already become addicted to this quite compulsive habit, to try growing a member of the family Proteaceae in their gardens. Obviously there are many parts of the world where attempting to grow the Proteaceae is out of the question, as they are basically a mild climate subject, but there are also many areas where it is only now being realised that it is possible to grow these glorious plants.

Lewis J. Matthews,
French Farm Bay,
Banks Peninsula,
New Zealand.

February 1993.

Zoë C

CONTENTS

◄ PLATE 2: *Leucospermum cordifolium* 'Harry Chittick' is a good example of the many outstanding cultivars of this species.

Zoë C

INTRODUCTION

In today's world of abbreviations and simplifications most members of the family Proteaceae are with increasing frequency being referred to simply as proteas, hence the title of this book, *Proteas of the World*. In fact, more correctly, the family Proteaceae is the subject.

The Proteaceae are an ancient family of plants found almost exclusively in the Southern Hemisphere. Scientists have traced the origins of the family back more than 300 million years to the time of the supercontinent, Gondwanaland, before it slowly broke up to form Africa, South America, Australia and the islands of New Guinea, New Zealand, etc. Even at that time the family had divided into its two distinct subfamilies, Proteoideae and Grevilleoideae. Today, members of this unique family are found in Australia (more than 800 species), South Africa (329 species), Central and South America (about 90), the islands east of New Guinea (80), New Caledonia (45) and a few in central Africa, south-east Asia, New Guinea and New Zealand (2).

When early explorers and botanists first sent examples of protea flowers back to Europe, they aroused tremendous excitement and interest and it was in the wake of this in 1735 that Carl Linnaeus, the famous Swedish botanist, named them *Protea* after the legendary Greek Sea God Proteus. It is considered unlikely that he realised just how appropriate his choice of name was. Greek mythology tells us that Proteus was able to change his form at will when pursued by those seeking the advice which he was so reluctant to give. It must have been confusing to early botanists to examine widely differing flowers which, although exhibiting the same basic characteristics, came from radically different plants. When the huge flowers and smooth leaves of the giant *Protea cynaroides* are compared to the deeply toothed leaves and toothbrush-like blooms of *Grevillea hookeriana* or if one looks at the prostrate, mat-forming habit of *Leucospermum prostratum* with its delicately perfumed little 'pompon' flowers and contrasts that with the intriguing Mountain Rose, *Protea nana*, bearing pendulous, mahogany-red bell-like blooms on slender branches clothed with needle-like foliage the variations in form are immediately apparent. Protea is indeed a remarkably appropriate name for the entire family.

The family Proteaceae contains over 1400 species in 60 genera. From a horticultural or home garden viewpoint, however, it is the hybridisation and selection of improved and significantly varying forms that is the most intriguing aspect of protea cultivation. Many species within genera, even in their natural habitats, are almost promiscuous in their readiness to cross-pollinate. Despite this, hybrids are not seen as often in the wild as in cultivation where individual species are usually grown in closer proximity than would be the case naturally. Over the last few decades people have taken a hand in assisting nature with the creation of many lovely hybrids, notably among the proteas, leucadendrons, leucospermums and grevilleas. Most horticulturists involved in this field consider that they are only touching the tip of the iceberg in hybridising the Proteaceae and the potential for future new cultivars is enormous. Certainly from the point of view of commercial cut-flower growers and florists it is the continual development of new and improved varieties which is so exciting. Already, in a relatively short space of time, the number of selections and hybrids runs into the hundreds, leading to the conclusion that ultimately there will be thousands to deal with. An international register of protea varieties has been proposed with the aim of compiling a register of names to avoid duplication and to record full descriptions. This has also been deemed necessary to ensure that only sufficiently distinctive varieties are given the status of registration.

Of equal significance to both horticulturists and home gardeners is the study of the requirements of the Proteaceae for successful cultivation. It soon becomes obvious to the would-be protea grower that these requirements are very different from those of the average garden plant and it is essential that they be understood to achieve success. They are not complex and in fact add to the unique qualities of these beautiful plants. They are more fully discussed in the following chapter.

◀ PLATE 3: *Protea* 'Coleman's Hybrid'. An unusual *P. magnifica* seedling which displays many desirable habits for both home garden and commercial picking purposes. It is expected to become available in most protea growing countries.

THE CULTIVATION OF PROTEACEAE

RELATING SOIL AND CLIMATIC REQUIREMENTS TO THE NATURAL HABITAT

The successful cultivation of plants of the Proteaceae family, as with all plants, is to a large extent dependent on a fundamental understanding of their natural habitats. The Proteaceae are possibly more specific in their requirements than many plants, and yet their needs are indeed basic, for excessive care and attention in the form of fertiliser application or too much water often prove fatal. The reasons for this are clearly apparent once the natural growing conditions (habitat) are considered; soils that are very free-draining and usually markedly acidic in character and impoverished with notable mineral deficiencies. Climates are essentially Mediterranean — either cool and moist over winter with arid summers (for example, the mountains of the Cape in South Africa), or warm, semi-desert areas (the coastal sands of the Cape or the sand plains of Western Australia). The vast majority of species (with the exception of some of the Queensland and New Guinea examples) are not found in humid, tropical environments. Even the Chilean Fire Bush (*Embothrium*) is found growing in the cooler, high rainfall regions of the Andes Mountains.

NUTRIENT REQUIREMENTS

Understanding the proteoid root system, so much a part of scientific dissertation in relation to nutrient requirements of the Proteaceae, is an integral part of successful cultivation. Proteoid roots are a highly developed system of white, waxen rootlets which may be seen either as a dense cluster along lateral roots or as flattened extensions of main roots. During the main spring and autumn periods of growth and particularly in low nutrient soils, the proteoid rootlets are most conspicuous. This distinctive root system is believed to have evolved over millions of years for the purpose of efficiently seeking out both nutrients and moisture in poor, arid soils. It is further propounded that the Proteaceae evolved in this manner so that they could survive in comparatively hostile environments where the likelihood of invasion by competitive species would be minimal.

Research work carried out by a number of scientific establishments (probably first in 1974 by Hocking & Thomas, Lincoln College, New Zealand, in *Evolution of the Proteaceae and Cultural Implications*) has shown conclusively that heavy losses in Proteaceae result from applications of both phosphorus and calcium. This finding had the interesting spin-off of superphosphate being topdressed over invasions of *Hakea* species on agricultural land in Australia to assist in eliminating them. As a general rule, proteoid rootlets do not develop in rich soils. Plants may grow satisfactorily, or even lushly, temporarily, until the stress of dry conditions when the lack of proteoid rootlets to seek out moisture may cause the plants to collapse.

The foregoing leads to the reasonably safe assumption that fertilisers should not be applied to proteaceous plants. There will be exceptions, however, and growers with very poor soils have found that light applications of fertiliser can be beneficial, in particular nitrogen in lighter, sandy soils. This applies also to container growing of the plants by nurserymen. The medium used by modern nurseries is invariably a sterile blend of peat and pumice, bark, sand, etc., and it is necessary to use slow-release fertilisers to give some nutrients. However, nurseries always use the fertilisers which have the lowest phosphorus content. Dolomite and trace elements, in particular iron in coated form, are also often added to the medium and individual mixes are now usually tested before use to determine the precise levels of nutrients available to the young plants. Some genera and species are more sensitive than others to fertilisers and nurserymen have realised the importance of using a range of media with varying nutrient levels. Many of the Western Australia *Banksia* species and some species of *Protea* are much more sensitive to high concentrations of nutrients than are most leucadendron and leucospermum species. Certain protea, *P. nana* for instance, when container grown will require only very small amounts of fertiliser and will quickly succumb to heavier levels which may well be ideal for other species or genera of Proteaceae.

SOIL TYPES

From the above it can be seen that soils that have a low nutrient level are the most desirable. A low pH level, indicating an acidic nature, of between pH 5 and pH 6 (pH 7 is neutral) is also desirable, although a lower figure is acceptable as some proteas occur naturally in soils with a pH of as low as 3. Soils that are naturally alkaline (pH 7 and higher) are very difficult to make acid (unlike the reverse, where the adding of lime is a simple remedy), and for the determined protea grower it is a more practical solution to grow those species which are lime tolerant.

Drainage cannot be over-emphasised as one of the essential criteria for success in protea cultivation. With only a very few exceptions, the Proteaceae require well-drained soils. Many species enjoy good supplies of water, particularly over the winter months in the case of those native to many parts of South Africa's Cape. For the plants to take advantage of natural or artificial water, it must drain away freely. Poor drainage inevitably results in root rot and the various associated fungal pathogens such as *Phytophthora*. A theory has also been advanced that proteoid roots produce toxins which need to be washed away by water passing freely through the surrounding soil. Soils which are heavy in character may prove suitable for growing proteas provided the ground slopes sufficiently to allow water run off. Flat situations where damp soils prevail may be made suitable either through raised bed plantings or digging drainage ditches. Some of the best soil types for growing proteas are sandy loams, gravelly soils, silts overlying stones and those with a sandstone subsoil. In summary, therefore, the best soils are those that drain freely, are acid (with a pH of less than 6.5 ideally) and are low in nutrients (especially phosphate).

WATER

Proteas do require adequate moisture. It is a fallacy to assume that because well-drained soils are necessary for success that proteas 'like it dry' or are necessarily drought resistant. Certainly once thoroughly established they can be drought tolerant, although even this tolerance is dependent on soil types. An extremely free draining, light or sandy soil will need water applications long before one with a clay or sandstone subsoil. A great many of the proteas that are so well known in cultivation have their origins in South Africa's Cape where winter rain carrying on through spring is the norm. By December or January rainfall is becoming scarce and dry autumns are a regular occurrence. It is for this reason that many commercial growers who irrigate through spring and early summer to encourage maximum growth and stem length will 'wean off' the water supply from January or February (Southern Hemisphere) to simulate natural conditions. Indeed, excess water applications to species such as *Protea magnifica* during February can prove fatal.

As with many ornamental plants, overhead watering, especially during bright sunlight hours, is not recommended. For many reasons the best method of irrigation is the drip system.

In areas of limestone deposits where ground water is notably alkaline it may pay to restrict irrigating to rainwater supplies, or at least ascertain that the pH of the groundwater is not significantly higher than pH 7 (neutral).

SITUATION

Although not always possible, it can often be beneficial to research a plant's natural habitat before selecting the position for planting. Habitats can vary greatly, ranging from sandy, mild, coastal situations to rocky, cold, mountain locations or even occasionally places which become boggy in winter. In spite of this it is fairly safe to generalise. Reasonably sunny, open, airy situations are best for most species and varieties. Sunny positions usually enable the flowers or bracts to develop their optimum colour; shaded positions make for softer colours and weaker, spindly plants. Open and airy positions allow plants to grow to their best natural shape and let air circulate and help to keep foliage free from fungal problems.

SHELTER

Shelter is not often a significant consideration for home gardeners as most proteas will tolerate some wind, but strong gales can damage plants and in more exposed situations some form of shelter to filter wind may be advisable. Solid shelter is undesirable as this can cause a 'dumping' or 'pancaking' effect by the wind which is more damaging. The old idea of a few, large, strategically placed boulders around the base of a shrub is an excellent method of preventing rocking (and resultant root damage). They also serve as a mulch and keep the ground cool.

PRUNING

Many of the Proteaceae benefit markedly from a regular, yearly pruning. It is most important with many *Leucadendron* species and hybrids and also with *Telopea*. The degree of severity of pruning will vary according to the age of the plant and with the species. Younger plants will often respond better to pruning than older ones. Species which have lignotuberous rootstocks, such as *Protea cynaroides*, *Telopea*, *Mimetes*, *Banksia* such as *B. candolleana*, *B. robur* and numerous others can benefit markedly from very hard cutting back. This has almost the same effect as the bush and veld fires to which they are often subjected in their natural state and causes a virtual regeneration with vigorous new shoots coming away from the base of the plant. Species which do not have such rootstocks may not respond well to hard pruning and it is as well to prune back only to within the previous season's growth.

In the case of proteas, removal of spent flowerheads is an essential minimal pruning and encourages more vigorous new growth and longer flower stems. Young plants should have any dominant 'leaders' pruned back to encourage compact, branching growth habits from an early stage — this basic formation of branch framework at a young stage of development is just as important as with fruit trees, for instance, in creating a well-shaped adult plant.

Leucadendrons, particularly the vigorous modern hybrids such as *L.* 'Safari Sunset' and *L.* 'Silvan Red', should have most of the current season's growth removed as if it was being picked for cut flower purposes. Any thick, older wood from the centre of the bush should be thinned out at the same time. This should be done in early spring before the onset of new growth.

Telopeas, more than with almost any other member of the family, respond to hard pruning. This should be done in late spring, as the flowers are finishing. New shoots will spring from within the finishing bloom if this is not carried out. A total thinning process should be carried out at this time, with a chainsaw if necessary, to encourage as much new growth from as low down on the bush as possible. If this is not done, a tall 'leggy' shrub will result on which the blooms are high up, almost out of sight and virtually unpickable.

Leucospermums generally require little pruning. Some shaping, removal of dead flowers and a healthy picking are normally sufficient to retain a well-shaped plant.

Banksias generally respond in much the same way as leucadendrons to vigorous pruning while young, shooting away well to create a more branched, bushy and compact shrub. Those species having lignotuberous roots may be cut back much more severely than others. Before heavy pruning is carried out it is as well to ascertain which growth form the plant has.

FROST

Frost is one of the most difficult subjects on which to generalise in relation to the cultivation of proteas. Many factors can influence the degree of damage inflicted by freezing temperatures and guides can only be in the most general terms. One of the certainties is that in regions experiencing heavier frosts only the hardiest species and cultivars should be attempted. These are usually those which have their origins at higher altitudes. *Protea venusta*, for instance, can occur at altitudes of up to 2000 m where winter snow is commonplace. *P. effusa*, *P. aristata*, *P. subvestita*, *P. amplexicaulis* and *P. sulphurea* are from similar habitats which experience cold winters. Species such as these have proven frost tolerant in many gardens where frost is a problem. However, even these species, if grown in soils where nutrient levels are high and lush autumn growth occurs, may be damaged by early frosts. Plants that are slow growing through hard, poorer conditions will invariably prove more frost tolerant.

It is very important to protect young, newly transplanted plants from frost damage. An easy solution is to cover a plant with a sturdy carton overnight if heavy frost threatens, but there are many alternatives to this simple method and some gardeners use semi-permanent structures of posts and scrim or some of the recently developed 'frost cloth' materials to suit their particular requirements. Whatever method you choose, in colder climates tender young plants should always be protected until they have thoroughly acclimatised and gained some size.

Locating plants in a position with a western aspect or where the early morning sun's rays cannot amplify the effects of frost on plants has worked well for many gardeners. Similarly, hosing frost off foliage before the sun reaches the plants can minimise damage.

PESTS AND DISEASES

Like all plants, the family Proteaceae is subject to its share of pests and diseases. Obviously, the number and type of insect pests encountered will vary considerably from one country or region to another, but fungous diseases know few boundaries and are found to some extent in most places.

Hygienic growing conditions — well drained soil, space between plants to allow air circulation, light and open situations, and prompt removal and disposal of dead or diseased plant material — reduce the likelihood of pests and diseases occurring. Prevention is always better than cure. However, even if you follow these basic rules of prevention, problems can still occur. Obviously they will be easier to manage if they are tackled at the first signs and not left to reach epidemic proportions. At the first sign of pest or disease remedial action should be taken. For many commercial flower growers a regular, preventative spray programme will be a way of life, whether problems are evident or not. Particularly with growers who export protea blooms, elimination of all traces of pests or diseases is essential to retain both access to and good returns from offshore markets. In the monoculture environment created by the commercial growing of perhaps thousands of one species or variety, problems can rapidly multiply to crisis proportions hence the necessity for preventative measures.

Whether the problem is leaf spotting or dieback, *Phytophthora* root rot, caterpillar or beetle damage to flowers or bracts, the pests and the cures may vary from one country to another and both gardeners and commercial growers are recommended to seek professional advice from qualified personnel in their region.

WEED CONTROL

As with all ornamental plants, proteas respond to the minimising of competition from other species, be they weeds, grasses or ornamentals. Allowing grass to grow close to the plants, as in a lawn, can have the beneficial effect of keeping the roots cool and undisturbed, but allowing grass or weeds to run rampant can take too much moisture and nutrients from the soil. Also, the choking effect of grass and weeds amongst plants can lead to the onset of fungous disease, cause misshapen bushes and limit new growth. Control is very necessary.

Commercial growers have used herbicides, both pre-emergent and post-emergent, for many years amongst proteas with little damage being recorded. With both contact and systemic (absorbed through the sapstream) weed sprays, types should be used which leave no residues in the soil. There are many of these available under different proprietary names. Obviously pre-emergent sprays leave residues for a limited period. With all herbicides it pays to follow the directions implicitly — incorrectly concentrated mixtures are not only costly and wasteful, but may cause damage to the plants. Excessive spraying over long periods has been found to have harmful build-up effects, more so in certain soil types, and caution should always be exercised.

Mulches have been tried over the years but to date there have only been two satisfactory and generally available types. In the home garden where appearance is paramount and smaller areas reduce costs, coarse pine bark has proved its worth. In the commercial sector the woven, polypropylene mesh often referred to as weed matting has achieved general acceptance. The weave allows some moisture and air penetration and does not sour the soil as with solid polyethylene film. The habit of some gardeners of heaping grass clippings around trunks can be extremely harmful, causing rotting of the bark.

Hand hoeing, while obviously uneconomic in commercial plantings, is usually quite effective in the garden, and can have the beneficial effect of creating a dust mulch, provided the practice is restricted to light, surface hoeing only. Deeper activity can seriously damage the upper surface roots, and it should be remembered that as a rule the important proteoid roots are concentrated near the surface.

PROPAGATION

Most members of the Proteaceae can be propagated vegetatively, through the taking of cuttings, by budding or grafting (though to a lesser extent), and by the sowing of seed. Some examples and relevant comments follow.

Banksia While certain species such as *B. ericifolia* and *B. integrifolia* may be propagated vegetatively, most nurserymen have found that seed is the easiest and quickest method. This is best sown fresh in spring or autumn in a free-draining medium such as peat and pumice, sand and perlite or similar. Some variation may be detected in the plants produced by this method but it is not usually significant.

Leucadendron With one or two exceptions (notably *L. argenteum*) most species and varieties of leucadendron are propagated from cuttings. Leucadendrons may be quite variable when grown from seed, the progeny not being identifiable as a particular form or sex until it has flowered. Because of this, and also to ensure hybrids and selections remain true to type, vegetative propagation has become essential in the nursery and cut-flower industries.

Protea The foregoing comments also apply to a large extent to proteas, except that many do come true from seed. Due to the wide variation among seedlings of *P. neriifolia*, *P. longifolia*, and *P. laurifolia* these species are not usually grown from seed for nursery sales and selected forms are grown from cuttings. Species such as *P. nana*, *P. pityphylla* and *P. effusa* rarely show any departure from the normal characteristics and are routinely propagated from seed as well as cuttings.

Leucospermum As a general rule, leucospermums are now propagated vegetatively, cutting-grown plants more speedily reaching a saleable size and usually producing a more robust specimen which flowers within two years. Some of the more unusual or rarer species are grown from seed, largely due to lack of stock plants available to the nurseryman for gathering cutting material. Propagating material is usually gathered in late summer, the cuttings selected usually being on the softer side of semi-hard. Heating cables, heated pads or a glasshouse to provide warmth will give quicker rooting of cuttings.

Vegetative propagation Cuttings are generally taken of semi-hardwood material in the autumn (except in the case of leucospermums — see above). Material is usually up to 20 cm long, and has the lower one third of foliage carefully stripped or cut off flush with the stem. A hormone treatment is given to encourage the formation of roots. This is usually either a powder, such as 'Seradix' No. 2, or liquid IBA (Indolebutyric Acid), in varying strengths. The cuttings are normally placed in their tubes or trays in a sheltered, semi-shaded position where they are not subjected to excessive heat or humidity and the foliage occasionally misted to prevent transpiration during the rooting period. Rooting may take a few weeks or several months according to conditions and the type of material used. In the case of some varieties the use of 'bottom heat' (soil heating cables or mats) can speed up the process.

Tissue culture In recent years much research has been done into the tissue culture or micro-propagation of proteaceous plants. Limited success has been achieved with telopeas and grevilleas in Australia and New Zealand, but to date only a few results have been good enough to enable the commercial tissue culture laboratories to undertake full scale production of some hybrids and selections. When results improve, the commercial flower-growing sector in particular should benefit with quicker 'bulking up' of new and improved varieties.

Budding and grafting While budding and grafting are not practised widely by nurserymen in the day-to-day propagation of Proteaceae they do have important applications. In areas where specific species are difficult to grow due to excessive soil moisture or the presence of soil-borne fungal diseases, the use of rootstocks which are resistant to these problems can enable more difficult species to be grown successfully. But these methods are more suited to the keen enthusiast or determined commercial flower grower. Production of grafted and budded Proteaceae in the nursery is often too costly in time and labour to be economical.

In Australia some delightful results have been achieved with the grafting of grevilleas. In particular, the use of *Grevillea robusta* as a standard stock in creating weeping, standard specimens of *G.* 'Royal Mantle' (a prostrate, ground cover hybrid) and other similar species and hybrids has proven highly successful.

USES

The Proteaceae have a broad range of uses both in landscaping and for floral decoration. To a lesser extent, and possibly more of historical interest, are the uses of the timber, in particular by the pioneer settlers of both South Africa and Australia. Several species of banksia were noted for their qualities as firewood, *Banksia menziesii* being commonly referred to as 'Firewood Banksia', while others have been used in cabinet making. In South Africa, the 'Waboom', *Protea nitida*, was used in the making of wagon wheels and for furniture. Its bark was also used for making tannin and ink. *P. repens*, the 'Sugar Bush', so named for its copious supplies of nectar, was used to make a syrup known by Cape colonists as 'Bossiestroop'. This mixture was used initially as a sweetener and later for curing coughs and other chest complaints. In New Zealand, the attractive and distinctively grained timber of *Knightia excelsa*, known as 'Rewarewa' or 'Honeysuckle Tree' has been used in the manufacture of parquet flooring and for inlaid work. *Grevillea robusta*, the Silky Oak, so named for the appearance of its timber, has in the past been widely used in its native Australia for purposes ranging from cabinet work to coachbuilding and even cooperage.

For some gardeners the appeal of many proteaceous plants will lie in their attractiveness to nectar-seeking birds. In South Africa, Australia and New Zealand, the many native honey eaters are enthusiastic visitors to the nectar-producing flowers.

The garden landscape In many gardens, especially those with soils which are well drained but low in nutrients, provided the climate is suitable, the Proteaceae can become a feature; the vast array of types and different genera blending together to perfection. Taller growing species may be used for background effect, a foil for other plants or for shelter. Shrubbier, more typical examples can be used for colourful display and picking at varying times of the year. So many proteas and leucadendrons give autumn and winter colour, while examples such as the telopeas provide a spectacular sight in spring. Leucospermums give the early to mid-summer display to complete the year-round flowering which is so easy to achieve with this spectacular family of plants.

For low-growing specimens to use in the foreground of shrub borders or in raised planters there are a multitude of smaller species and varieties from the proteas, leucadendrons, leucospermums, grevilleas and other genera. There are also many prostrate examples to use for ground cover or to spill down a bank or over walls.

Floral art For the keen floral artist or amateur home flower arranger, the flowers of the Proteaceae are popular for their versatility, variety, long vase life and availability in times of shortage of other flowers. Large protea blooms suit big, dramatic arrangements, while long-stemmed leucadendrons give line and rich colour. Small proteas and leucadendrons are ideal for posies and table pieces, banksias and waratahs for sheer impact, or leucospermums for summer colour — the possibilities are almost endless.

Commercial flower growing and floristry The growing of proteaceous plants for the world flower markets has come into its own since the 1970s. Prior to that most material available was either picked from wild stands, mainly in South Africa's Cape and to a much lesser extent in Western Australia. The number of commercial growers involved in setting up plantations was very small, and it is only since the late 1970s and early 1980s that numbers have increased rapidly and growers can even be found in such previously unthought of places as Maui in the Hawaiian Islands, Southern California and even Israel. The three main producer countries are South Africa, Australia and New Zealand.

Florists in many countries have been slow to realise the potential of proteas, largely through inefficient marketing of the product. Growers see this as a challenge and are working on various methods of promotion in different markets. Some florists see proteas as simply a cut flower to sell by the bunch or individually, or maybe to use only in arrangements in large offices and the foyers of commercial buildings. Others use very large quantities of leucadendrons and smaller proteas in most of their pieces. Almost everywhere the uses of the flowers will vary according to the tastes and requirements of the individual florist, making the task of promotion even more difficult!

It appears that the main attributes to promote are the long vase life and dramatic, unique appearance of the blooms. The future of the protea flower industry will probably hinge on a professional, unified and intensive approach to marketing. It is widely believed also that an integral part of this marketing strategy will be stringent quality controls so that the flowers achieve a reputation around the world for both beauty and perfection as well as longevity.

THE FUTURE

Development continues with the selection of cultivars which exhibit more desirable characteristics — longer vase life, no leaf blackening, longer stems, brighter colours, more disease resistance, and better travelling qualities for cut blooms, for example (just some of the qualities that flower growers, nurserymen, florists and home gardeners are always looking for). It is to the industry's credit that many of these qualities are being found through careful observation and selection and scientific research.

Hand-in-hand with developments and selections is the continuing research into improved propagating methods. It is quite possible that once tissue culture techniques are perfected this will become the main source of supply for new varieties. This should significantly speed up the process of getting commercial quantities to growers for planting out.

The future for the Proteaceae is full of promise; they are being better understood, their popularity is increasing and new developments are promoting both these trends.

THE PROTEACEAE OF AUSTRALIA

The family Proteaceae is more strongly represented in Australia than in any other country, with 42 genera and about 860 species. South Africa, by comparison, has the next largest number, with only 13 genera and 329 species.

It is beyond the scope of this book to illustrate and describe all the amazingly diverse species of the Proteaceae; instead the better known and significant species have been selected for publication. The following list covers the genera found in Australia and gives the number of species. With some, *Lomatia* for instance, there is an overlap with South America, where other species are recognised.

Adenanthos; 32 species, 4 subspecies.
Agastachys; 1 species.
Alloxylon; 3 species.
Athertonia; 1 species.
Austromuellera; 1 species.
Banksia; 75 species, 13 varieties, 1 subspecies.
Bellendena; 1 species.
Buckinghamia; 2 species.
Cardwellia; 1 species.
Carnarvonia; 2 species.
Cenarrhenes; 1 species.
Conospermum; 36 species, 2 subspecies.
Darlingiana; 2 species.
Dryandra; 57 species.
Floydia; 1 species.
Franklandia; 2 species.
Grevillea; 273 species, 61 subspecies.
Hakea; 130 species.
Helicia; 9 species.
Hicksbeachia; 2 species.
Hollandaea; 1 species.
Isopogon; 34 species, 3 varieties.
Lambertia; 11 species.
Lomatia; 9 species.
Macadamia; 7 species.
Musgravea; 2 species.
Neorites; 1 species.
Opisthiolepsis; 1 species.
Orites; 8 species.
Persoonia; 54 species.
Petrophile; 39 species.
Placospermum; 1 species.
Sphalmium; 1 species.
Stenocarpus; 7 species.
Stirlingia; 3 species, 2 varieties.
Strangea; 3 species.
Symphionema; 2 species.
Synaphea; 9 species.
Telopea; 4 species.
Triunia; 4 species.
Turrillia; 1 species.
Xylomelum; 5 species.

BANKSIA BAXTERI

Commonly known as Bird's Nest Banksia or Baxter's Banksia. Collected in 1829 by William Baxter and named for him. Natural distribution is a broad area near the south coast of Western Australia, on sand plains and frequently in association with *B. coccinea*, the Scarlet Banksia.

DESCRIPTION:
Growth habit: densely erect until maturity at 8 to 10 years when it develops into a wide, dome-shaped shrub 3 m high.
Leaves: long (to 170 mm), particularly attractive, very deeply serrated to form almost 'V'-like sections, being divided to the midrib. The ornamental, soft, young growth is bright, rusty-brown with reddish stems.
Flowers: the erectly held, terminal flowerheads on particularly long, straight stems are 80 to 100 mm wide, dome-shaped, an attractive shade of lemon yellow when fully open, and an appealing lime green at the bud and half-open stage. The main flowering period is from summer through to early autumn.
Variants: this is a very distinct species and throughout its distribution range shows virtually no variations apart from size of leaves.

CULTIVATION:
Regarded as an easy banksia to cultivate in areas where soils are very free-draining, such as deep sands, silts overlying gravel, etc. Situations with heavy clay subsoils are usually unsuitable and losses are frequent. Acid to neutral soils are all suitable, but alkaline ones are not. Open sunny positions are preferred, and some protection from frost will be necessary, particularly while young plants are becoming established. Once established, *B. baxteri* is wind tolerant and is even recommended as a windbreak in parts of Australia. Some light pruning will encourage branching, but picking of the blooms will serve the same purpose.

PROPAGATION:
Normally propagated from reasonably fresh seed, sown in spring or autumn in a free-draining medium of peat and sand, pumice, perlite, etc. Most propagators have their own special blend.

USES:
An especially versatile banksia, being suitable for shelter, ornamental planting or commercial cut-flower production. Possibly its greatest potential is as a cut flower; the handsome, exquisitely coloured blooms are produced on the much favoured long stems, have an excellent vase life and are often dried very successfully. The leaves are also treated in this way by florists and floral artists.

PLATE 4: *Banksia baxteri*.

Zoë C

BANKSIA BURDETTII

Referred to commonly as Burdett's Banksia. Named after W. Burdett, a South Australian wildflower grower who first collected material of this species north of Perth around 1930.

Occurs naturally in a restricted belt in the deep sands of the northern sand plains of Western Australia. Considered to be an endangered species due to the encroachment of agriculture.

DESCRIPTION:

Growth Habit: usually fast growing in favourable conditions, reaching a maximum of 4 m, with a dense, bushy habit, often as wide as it is high. Under cultivation, however, it is rare to see a specimen of much more than 2 m x 2 m.

Leaves: 160 to 200 mm in length, up to 20 mm wide, lightly serrated, slightly twisted.

Flowers: particularly ornamental and sought after as cut flowers. Of the typical 'acorn' shape when fully developed, they commence as an intriguing woolly, greyish-white bud and become a brilliant orange as the styles open out. Up to 120 mm long and 80 mm in diameter, the blooms are terminal and borne on stems of useful length. The main flowering season is late summer into early autumn, with some blooms occurring at other times.

Variants: no variation is recorded, largely due to the very limited distribution of this endangered species.

CULTIVATION:

Relatively easy to grow in the home garden provided an open, well-drained site is provided. Much slower growing and short-lived in heavier soils. Tolerant of extended drought conditions once established. Prune lightly within foliage area of stem to promote compact, bushy habit. Application of fertilisers, as with most of the Proteaceae, should be avoided, phosphates, potassium, nitrates and lime all being particularly harmful.

PROPAGATION:

Fresh seed usually germinates readily and seedlings reach flowering within 3 to 4 years.

USES:

An attractive, bushy shrub for the home garden and sought after as a cut-flower species. The only drawback is its relatively short flowering season; usually only a few weeks, but the beauty of the spectacular blooms more than makes up for this. It is grown as a commercial cut-flower crop in both Australia and Hawaii.

PLATE 5: *Banksia burdettii*.

ZoëC

BANKSIA COCCINEA

Variously known by the common names of Scarlet Banksia, Waratah Banksia and Albany Banksia, which serve to illustrate both its popularity and wide distribution. Found in a variety of soil types, from deep sands to gravelly and even marshy situations, throughout south-western Australia from Albany to the Stirling Range and east to the Young River.

DESCRIPTION:
Growth habit: generally a narrowly erect, slim shrub or small tree with several main upright stems. Height may vary from 4 to 8 m, but under cultivation 3 to 4 m appears the usual, particularly with picking of blooms or pruning to encourage a more branched habit.
Leaves: leathery, short, broad and with blunt ends, the foliage is bordered with small, prickly teeth which vary in occurrence and is deep green in colour with grey or white undersides. The leaves are quite distinctive, unlike any other banksia.
Flowers: equally distinctive, the spectacular terminal blooms are held stiffly erect on long, thick, rigid stems. They are squat and cylindrical in shape and may vary in size from 60 to 120 mm long with a diameter of anything from 60 to 150 mm. The colour of the beautiful blooms which open from furry, grey buds is also variable, ranging from orange through scarlet to dark red. The scarlet-red colouring illustrated is the more common shade. The flowering season is variable and extended, occurring from mid winter to mid summer with a peak in the spring months. Individual plants will frequently produce blooms over a period of 2 to 3 months.
Variants: as already mentioned above, the flowers are most variable, both in colouring and size. Foliage also is likely to exhibit size variation from one plant to another and the amount of teeth can be quite noticeable on the leaves of one bush and almost absent on another.

CULTIVATION:
Soils should be well drained and acid and it should be borne in mind that as this species is susceptible to *Phytophthora* (the fatal root fungus disease) free-draining soil is of paramount importance.

Sunny, reasonably open positions are desirable to encourage the development of well-shaped plants and good colouring of blooms.

B. coccinea has proven to be reasonably frost tolerant in many areas, established plants having withstood the occasional drop in temperature to -5 °C.

Pruning is beneficial to encourage multiple branching and certainly results in a significant increase in the number of flowers produced. It should be carried out judiciously, no cuts being made below green foliage.

PROPAGATION:
Increase is from reasonably fresh seed sown in spring or autumn.

USES:
A very popular species for home garden display, its neat, erect shape and response to careful pruning enables it to be accommodated in even small gardens. A lovely cut flower.

PLATE 6: *Banksia coccinea.*

BANKSIA ERICIFOLIA

Commonly known as Heath-leaved Banksia, an obvious reference to the erica-like foliage. Reputed to be the first specimen collected by Sir Joseph Banks at Botany Bay in 1770. Most colourful of the eastern Australian banksias. Widely distributed in sandy, coastal areas of New South Wales.

DESCRIPTION:
Growth habit: a compact, dense, bushy shrub to 6 m or more, often with branches to ground level. Sometimes assumes the habit of a small tree with age.
Leaves: masses of small, dark green, erica-like leaves, each very narrow and no more than 15 to 20 mm long, are one of the strong identifying characteristics of this species.
Flowers: varying in colour from soft bronze-orange to almost scarlet-red. Tall and cylindrical in shape, 200 to 300 mm long and 50 mm wide at flowering. Although the blooms may be partially hidden by leaf growth they are nonetheless a colourful sight. The main flowering season is through winter from late autumn to early spring.
Variants: apart from natural variation of flower colour, there are two varieties which are particularly distinct through foliage differences as well as geographical location. One, var. *macrantha*, has much narrower, more crowded leaves and deeper coloured flowers. Hybrids with *B. spinulosa* are recorded, including one with almost black foliage.

CULTIVATION:
Experience has shown this species to be adaptable under cultivation to light and heavy soils and sun or semi-shade. It is also tolerant of frost to approximately -5 °C once established. However, ideal soils are well drained and acid, and sunny situations usually give better specimens.

Light pruning within green foliage will promote new growth resulting in bushier habits.

PROPAGATION:
B. ericifolia may be propagated from semi-hardwood cuttings, usually taken in autumn, and also from reasonably fresh seed sown in the autumn.

USES:
Well known in Southern Hemisphere gardens, particularly the eastern states of Australia and throughout New Zealand. Regarded in both countries as one of the easiest banksias to cultivate. Provides a dense, attractive, larger shrub suitable as shelter, for background effect with its deep green foliage, or as a specimen with its neat habit and colourful, large flowers. It has uses for lower-growing shelter on farms as the foliage is not palatable to stock. It has the added bonus of being particularly attractive to nectar-seeking birds. It is grown as a cut-flower crop in Hawaii on the island of Maui.

PLATE 7: *Banksia ericifolia.*

Zoë C

BANKSIA GARDNERI

Formerly known as *B. prostrata*. First collected in 1823 by William Baxter. Found growing in sandy soils in shrubland country on Western Australia's south coast. Named after C.A. Gardner, government botanist in Western Australia from 1929 to 1960.

DESCRIPTION:
Growth habit: one of the few, but intriguing, prostrate species of banksia, *B. gardneri* has a very low habit with the branches flattened against the ground and the leaves held erectly above them.
Leaves: erect, rigid and deeply lobed almost to the midrib, the medium green leaves may measure 100 to 400 mm long and 20 to 60 mm wide.
Flowers: the conspicuous rusty-brown flower spikes are borne at the end of the creeping branches and held stiffly erect. Cylindrical in shape and measuring up to 120 mm long and 60 mm wide when fully developed. The main flowering season is over spring and early summer.
Variants: there are three distinct varieties of *B. gardneri*, distinguished from each other by significant variations in leaf form, less obvious but still distinct differences in flower colouring and considerable variations in flowering time. A separate but related species is *B. goodii*. There are four other species of prostrate banksia, *B. repens* probably being the best known of them.

CULTIVATION:
Slow growing, but a reliable and relatively trouble-free species, apart from in some eastern Australian regions, *B. gardneri* has a lignotuberous rootstock from which it springs away into new growth after bush fires. Because of this it may be severely pruned if necessary. Regarded as an interesting rock garden subject under cultivation, it requires well-drained sandy or gravelly soils to grow well. They may be either neutral or acid in reaction. Full sun or partial shade are both suitable situations and it will tolerate light frosts once established.

PROPAGATION:
This species is always propagated from seed.

USES:
As with any of the prostrate banksias, this species is a fascinating addition to rock gardens, the foreground of shrub borders and similar applications. Even when not in flower, the foliage effect is unique and never fails to arouse attention among gardeners. It has no picking potential as a cut flower.

PLATE 8: *Banksia gardneri*.

Zoë C

BANKSIA GRANDIS

Bull Banksia is the more popular common name, sometimes also referred to as Giant Banksia. The specific name comes from the Latin *grandis* — great, probably referring to the large leaves although the flowers are also very large. A common species throughout the south-western coastal region of Western Australia.

DESCRIPTION:

Growth habit: *B. grandis*, although a large, bushy shrub for the first four or five years of its life, soon assumes a definite small tree shape with a prominent trunk. It may reach a height of up to 10 m in favourable conditions. There are smaller-growing, coastal variants with spreading habits — see under 'Variants', below.

Leaves: possibly the most dramatic of any banksia, the leaves of *B. grandis* may be up to 500 mm in length, are a shining dark green with grey undersides and are deeply divided, almost to the midrib, giving the appearance of large, jagged teeth. They form a circular collar around the flower spikes. Young leaf growth is particularly attractive, being a pale bronze or red colour and soft and woolly in appearance.

Flowers: large and dramatic, the flower spikes (inflorescences) resemble huge golden candles. They are borne on stout stems of medium length, are evenly cylindrical in shape, about 100 mm in diameter and up to 400 mm long although more usually 250 to 300 mm. The colouring is a soft sulphur yellow. The main flowering season is through spring into early summer.

Variants: a generally variable species in terms of flower and foliage size but also a low, spreading, or shrubby form from the south coast of the geographic distribution is recognised. This has possibly evolved because of the growth-stunting effects of the strong coastal winds.

CULTIVATION:

Although this is regarded as a difficult species to maintain under cultivation in the eastern states of Australia, it is an easy species to grow in New Zealand and several other countries, provided the basic requirements are present — good drainage, acid to neutral soil and adequate but not excessive moisture. Either full sun or part shade are suitable. Established plants have withstood winter frosts of -4°C on the author's property.

PROPAGATION:

Fresh seed germinates easily and seedlings will usually produce flowers within 4 to 5 years of planting, although some growers have reported plants taking 10 years to produce the first blooms.

USES:

The usual form cultivated is the taller type, and this makes a handsome small tree for background or specimen use. While its flower spikes are not grown for commercial picking, being too large and lacking good vase life, the leaves are picked and dried for use by both florists and floral artists in dried arrangements. The cones are also used for turning on lathes to make candlesticks and other ornaments.

PLATE 9: *Banksia grandis*.

Zoë C

BANKSIA INTEGRIFOLIA

Widely referred to in Australia as the Coast Banksia and sometimes White Honeysuckle. Possibly the most widely cultivated banksia and also the most common species of the eastern Australian states, occurring as it does from Victoria to northern Queensland in varying forms. First collected and described by Joseph Banks in 1770. The Latin *integrifolia* translates to entire leaf, referring to the smooth leaf margins.

DESCRIPTION:
Growth habit: one of the largest-growing banksias, reaching a maximum of 25 m, *B. integrifolia* develops a large trunk to 60 cm in diameter with the crown somewhat open in form. In exposed coastal situations, virtually to the water's edge in some instances, the harsh salt winds cause the trees to assume a more stunted and gnarled appearance.
Leaves: juvenile leaves on young plants are usually broad and irregularly toothed, but most foliage on mature trees is smooth margined (entire) and more lanceolate in shape. The dark green leaves have a white underside which tends to give the tree a silver appearance in wind. Mature leaves are usually up to 150 mm long and 20 mm wide.
Flowers: the cylindrical flower spikes are held erectly. They are usually terminal, but can even grow from the trunk on younger trees. Colour is a pale, silvery yellow, deepening to bronze with age. They may measure up to 150 mm long and 70 mm wide when open, although more usually are about 100 mm long and 60 mm wide. Some nectar-seeking birds are attracted to the blooms which are produced chiefly from autumn to spring with a few spikes appearing in summer.
Variants: there are three varieties of *B. integrifolia*, mainly distinguished by leaf dimension and geographical distribution though there is some variation in flower size and growth habits. A prostrate cultivar is available in Australia. This species is recognised as one of the most variable.

CULTIVATION:
A particularly adaptable banksia, tolerating light and heavy soils that may be acid or neutral, moderate frosts, strong winds and inland or coastal situations.

PROPAGATION:
May be grown from semi-hardwood cuttings taken in autumn or from the seeds which eject freely from the cones when ripe (unlike many banksia species which need fire to eject).

USES:
Widely used in both the eastern states of Australia and throughout New Zealand as a very ornamental shelter tree, in both exposed inland and coastal situations. Planted to attract nectar-seeking birds in some regions. Limited uses only for cut flowers due to lack of stem length. Dried seed cones used in ornaments and arrangements. Used increasingly as a rootstock on which to successfully establish difficult species.

PLATE 10: *Banksia integrifolia*.

ZoëC

BANKSIA LEMANNIANA

Named for a 19th-century English botanist. Sometimes known as Lemann's Banksia. Found growing abundantly in a wide area of Western Australia's south coast on stony sand or loam among other taller shrubs.

DESCRIPTION:
Growth habit: a fairly fast-growing shrub to a maximum of 5 m. Bushy and often spreading in habit when growing in open positions, more erect and sparsely branched when among other competing plants.
Leaves: deep green leaves have prickly, serrated margins and may be up to 100 mm long. The slightly wavy, but firm foliage is an attractive foil for the unusual flower spikes. The new growth is velvety and a striking, rusty-orange to red colouring.
Flowers: the freely produced and conspicuous flower spikes are pendulous, hanging like big lemon-yellow globes from the bush over spring and early summer. They usually measure 120 mm or more in length and 80 to 100 mm wide when open. Nectar is freely produced.
Variants: no significant variations to the type are recorded.

CULTIVATION:
B. lemanniana has proved adaptable to widely varying conditions. It has been grown successfully in a variety of soils including sand, gravel, loam and clay, provided drainage is good. It should be pruned lightly, but not below green leaf growth, to encourage bushiness. Tolerant of light frost once established.

PROPAGATION:
May be propagated from semi-ripe cuttings or seed.

USES:
Popular as a fast-growing screen or windbreak with the added bonus of the unusual, attractive, pendulous flower spikes. Also attractive to birds.

PLATE 11: *Banksia lemanniana.*

Zoë C

BANKSIA MARGINATA

Generally referred to as the Silver Banksia, due to the silvery appearance created by the white undersides of the leaves. A widely distributed species found throughout south-eastern Australia from South Australia through Tasmania and Victoria to New South Wales. The label *marginata* refers to the rolled-under leaf margins on some forms.

DESCRIPTION:
Growth habit: a very variable species, ranging from as little as 1 m in height in the case of the windswept coastal forms to as much as 12 m with the variants found in more sheltered environs. *B. marginata* usually assumes the habit of a much branched small tree or large shrub, but occasionally is seen as a more erect specimen. The small tree habit to 5 or 6 m is the more usual when grown in cultivation.
Leaves: foliage is quite narrow, usually less than 10 mm wide and varying from 20 to 100 mm in length. The quite bright green colouring of the upper surfaces contrasts with the undersides which are white tomentose; this is where the silver appearance comes from, particularly when foliage is ruffled by the wind.
Flowers: the cylindrical flower spikes are lemon yellow when open, slender and green in bud and may measure 40 to 100 mm long and up to 60 mm in diameter. The main flowering season is usually from late summer to mid winter, but may sometimes occur in spring. A free-flowering species, mature trees being spectacular when in full bloom.
Variants: a species that is as variable and as widespread as *B. integrifolia*. Some forms are low growing, others tall, the leaves can vary greatly, and some have lignotuberous rootstocks making them fire-resistant, while others do not. Natural hybrids with *B. integrifolia* have been observed.

CULTIVATION:
As with *B. integrifolia*, it is one of the easiest species to grow in garden conditions. It is adaptable to both well-drained and heavier soil types, and tolerates both very acid and slightly alkaline soils. It is usually frost tolerant to about -12 °C. Forms with lignotuberous rootstocks may be pruned hard to encourage compact, branched habits; others should be lightly pruned only.

PROPAGATION:
Usually grown from seed which germinates freely if reasonably fresh.

USES:
A particularly useful species for shelter, the thick habit creating an excellent windbreak. Ornamental appearance makes an appealing backdrop for other shrubs. Not grown as a cut flower.

PLATE 12: *Banksia marginata.*

BANKSIA MENZIESII

Firewood Banksia or Menzies Banksia are the usual common names. Named for an 18th-century English botanist. Found growing over a wide area of Western Australia, particularly from the Murchison River to Pinjarra.

DESCRIPTION:

Growth habit: a variable species, either forming a sturdy shrub of up to 3 m in height or an open, irregularly branched tree of 10 m. As it has a lignotuber it may be seen branching densely from the trunk or lignotuber after fire. In such cases growth habit usually assumes shrub form.
Leaves: the oblong leaves are a grey-green colouring, margined with small, prickly teeth, slightly wavy and may be up to 250 mm long and 40 mm wide, although more usually smaller than this.
Flowers: a particularly distinctive and striking feature of this species, the flower spikes are cylindrical when in bud but become acorn-shaped as they open. Their colours and arrangement are fascinating and like no other species. The bud commences as a lovely silver-grey shade, then becomes red-tipped with silver tonings. As the flowers begin to open from the base with the golden styles the whole spike becomes a sculpture of neatly arranged vertical rows of wine-red, silver and gold. The spikes or inflorescences as they are more correctly termed may measure up to 120 mm long and 80 mm wide when flowering. They are usually borne terminally on quite long stems over autumn and winter. The seed cones which follow are also strikingly different, carrying only a few large seed capsules which are irregularly spaced over a small, beautifully patterned cone.
Variants: heights and growth habit vary according to geographic distribution and the colouring of the flowers can range through pink shadings to almost entirely pale yellow.

CULTIVATION:

Not always an easy species to grow in cultivation, *B. menziesii* appears to need particularly well-drained soils, preferably deep sand, and requires sunny, reasonably sheltered positions. It has withstood frosts of -4 °C without damage. It can withstand and benefit from quite severe pruning.

PROPAGATION:

Always propagated from seed sown in autumn.

USES:

A particularly good cut-flower species, it is being grown in several parts of the world on a commercial scale for this purpose. As a garden plant, provided the conditions suit, it is a very lovely ornamental shrub to grow. Originally used by early settlers as a source of firewood, hence the common name.

PLATE 13: *Banksia menziesii.*

Zoë C.

BANKSIA OCCIDENTALIS

Red Swamp Banksia or Water Bush, so named for its usual habitat of damp swamp margins. Specific name is from the Latin *occidentalis* — western, probably referring to its Western Australia distribution. Found on the south-western coast of Western Australia from Augusta to Cape Arid.

DESCRIPTION:
Growth habit: of either shrub habit or small tree proportions to a maximum of 7 m in height. Normally quite erect and dense in habit.
Leaves: particularly narrow-linear, dull green on the upper surface, white tomentose on the underside. They are 50 to 150 mm long and only 2 to 3 mm wide.
Flowers: ornamental, cylindrical flower spikes, variable in size, 50 to 150 mm long and 60 to 90 mm wide when open. A shining, ruby red colour as a rule, sometimes paler. Mainly produced during late summer, the flowers can sometimes appear during winter and spring.
Variants: not particularly variable, this species has a dwarf form in a coastal population at the western end of its range which only grows to 1 m. This height variation may be due to the strong winds and to date it is not known if progeny have been grown to see if the characteristic persists away from its coastal habitat.

CULTIVATION:
A banksia that has proven easy to cultivate in many regions, but difficult in some. Generally a sunny position in either light or heavy soil is suitable provided some water is available over summer. Either neutral or mildly acid soils are suitable. This species may be pruned, but not below green foliage. Moderately frost hardy, it has tolerated temperatures of -8°C once acclimatised.

PROPAGATION:
B. occidentalis may be propagated easily from seed which may need firing to extract from the cone. This is one of the few banksia species which may also be propagated from semi-hardwood cuttings taken in autumn.

USES:
An ornamental small tree with uses for background effect, screening and shelter and attracting nectar-seeking birds. Versatile, being one of the only banksias to tolerate moist conditions. The attractive flower spikes may be picked, and although this species is not generally recognised for having commercial cut-flower potential it is grown for this purpose in Hawaii.

PLATE 14: *Banksia occidentalis.*

Zoë C

BANKSIA PILOSTYLIS

Not known by a common name, probably due to the fact that this species was first gathered and described in 1960 by Charles Gardner, government botanist of Western Australia 1929-60. The specific name refers to the soft styles of the flowers. Found growing in deep sand country in southern Western Australia, especially between Esperance and Ravensthorpe. The name is from the Latin *pilus*, meaning hair, and *stylus*, a style, referring to the hairy styles of the flowers.

DESCRIPTION:
Growth habit: a dense, multi-branched, bushy shrub, 3 to 4 m in height and frequently as wide as it is high.
Leaves: the narrow, dull green leaves have short, sharp teeth set well apart along the margins, are from 100 to 200 mm long and up to 20 mm wide, and have a wavy appearance. Young foliage is attractive, velvety golden-brown.
Flowers: the cylindrical, pale gold flower spikes open from soft, woolly buds and appear during late spring and summer. They usually measure 100 to 150 mm long with a diameter of 60 mm. Superficially the inflorescences are similar to those of *B. media*, to which it is related, but of a paler colour. They have an unusual mouse-like scent.
Variants: a very uniform species, the only variations occurring in terms of flower size and shading.

CULTIVATION:
Usually a fast-growing shrub in cultivation, *B. pilostylis* requires little attention, flowering within 3 years from seed. Slightly acid to neutral soils that are well drained, plus a sunny position are the main prerequisites for success. It is moderately frost hardy. Removal of old cones is normally sufficient pruning to maintain a compact habit. As with all the banksias, it is advisable to avoid the application of fertiliser.

PROPAGATION:
Propagated from seed sown when fairly fresh to be assured of viability. Autumn is the preferred time.

USES:
An ornamental species for the garden due to its compact, bushy habit and terminal, conspicuous flowers. Not suitable for picking purposes, partly due to lack of stem length, odour of the flowers and short vase life.

PLATE 15: *Banksia pilostylis.*

Zoë C

BANKSIA PRAEMORSA

Commonly known as Cut-leaf Banksia, probably referring to the blunt leaf ends, as does the Latin specific name which comes from *praemorsus* — bitten off. Its habitat is coastal dunes and cliffs near Albany on Western Australia's south coast.

DESCRIPTION:
Growth habit: a straggling and often sprawling shrub in its windswept coastal habitat. A fast-growing, dense, large, upright bush or shrub reaching 3 or 4 m in height under cultivation.
Leaves: the distinctive, lightly serrated leaves are deep green, blunt, up to 50 mm long and 10 to 20 mm wide. The comparatively small size of the leaves is in marked contrast to the large flower spikes.
Flowers: the long, cylindrical flower spikes are frequently a spectacular 300 mm or more in length and 80 to 100 mm in diameter when fully developed. Colouring is difficult to describe, but especially prior to full development there is a wine-red toning which gradually assumes more yellow as the styles become prominent. The main flowering season is from late winter through spring to early summer.
Variants: being a very localised species *B. praemorsa* shows little variation apart from the rarely seen yellow-flowered form.

CULTIVATION:
A generally easy banksia to grow provided the soil is light and preferably either stony or sandy to give perfect drainage. Well suited to coastal conditions and soils, it tolerates sand, sea winds and drought once established. Heavier clay soils inevitably result in losses. A soil with a fairly neutral reaction (pH 7) appears ideal, although mildly acidic soils are also suitable. While some gardeners have found this species frost tender, the author has succeeded with it in areas experiencing winter temperatures of -4°C without any sign of damage.

PROPAGATION:
Usually from seed, but some success can be achieved with semi-hardwood cuttings.

USES:
A very useful species, especially for sandy, coastal gardens, its wind and drought tolerance making it desirable. Good for shelter, its fast-growing and bulky habit and height of up to 4 m is also ideal for screening purposes. Flowers are also an attractive feature, their size and colouring against the foil of the small, neat foliage being quite remarkable. Not generally used as a cut flower, mainly due to the limited vase life and short stems, *B. praemorsa* is nevertheless used by floral artists in dramatic arrangements to very good effect.

PLATE 16: *Banksia praemorsa*.

BANKSIA PRIONOTES

Commonly referred to as Orange Banksia or Acorn Banksia. The specific Latin name refers to the saw-like leaf margins. First collected and described in the 1830s. A widely distributed species over a large part of Western Australia from the Albany area to the Murchison River. The name comes from the Greek *prionotus*, like a saw, referring to the leaf margins.

DESCRIPTION:
Growth habit: varying in habit and height from a large shrub or tree, 6 to 10 m in height, *B. prionotes* is a fast-growing species ultimately forming a handsome tree of erect form with a spreading crown.
Leaves: long, to 300 mm, and to 250 mm wide, the wavy, deeply toothed leaves are shiny, dull green and very characteristic of the species, particularly in terms of length.
Flowers: the 150 mm-long flower spikes are highly ornamental, assuming a pronounced acorn shape as they develop. Opening from a woolly, grey-white bud to become a bright orange colour, they are borne terminally on long, rigid stems. The main flowering time is from late summer through autumn to early spring.
Variants: natural variants occur throughout the distribution range, the main variations being in growth habit, from spreading shrubs to tall trees and in size of leaves.

CULTIVATION:
Regarded as a very adaptable species under cultivation, *B. prionotes* is fast growing, usually reaching flowering within 3 years, and tolerates both slightly alkaline and acid soils. Reasonably frost tolerant, mature bushes withstanding up to -5 °C, although young growth can be damaged by early or late frosts. Pruning back lightly, within green foliage, encourages branching.

PROPAGATION:
Always propagated from seed, which is best sown fresh in the autumn into a free-draining media, usually peat and perlite or similar.

USES:
A dramatic, colourful feature tree for the larger garden, this lovely species has been grown for cut-flower purposes, not only in Australia but also in Hawaii and Israel, but is generally considered too 'leafy', the foliage making excessive bulk in arrangements. As orange-flowered banksias for picking, the related *B. burdettii* and *B. hookeriana* take preference. It is one of the many banksias which are attractive to birds.

PLATE 17: *Banksia prionotes*.

BANKSIA ROBUR

Known by the common names of Swamp Banksia or Broad-leaved Banksia. The specific Latin name *robur* translates to strength or hardwood. Found growing in the swampy sandflats of eastern Australia, from New South Wales to Queensland, in areas subject to summer flooding.

DESCRIPTION:
Growth habit: a spreading shrub, branching from the lignotuberous rootstock, reaching a maximum height of 2 m or sometimes 3 m. If crowded by other shrubs it may assume a more slender, erect habit.
Leaves: the foliage of this species is a particularly distinctive feature. Frequently as long as 300 mm, and 120 mm or more in width, the leaves are stiff, often slightly wavy and lightly serrated along the margins. Colouring is a dark, glossy green on the upper surface and frequently a rusty tomentose on the underside. A prominent orange or yellow midrib is also a dominant feature. Young growth is a most attractive reddish-brown colour and very velvety in texture.
Flowers: at the late bud stage the spikes are a distinctive metallic blue-green, almost a peacock feather shading. As the flowers open they become a yellowish-green colouring. Large in size, frequently exceeding 150 mm in length and 60 to 70 mm in diameter, they are basically cylindrical in shape. The flowering period is variable, but predominantly over autumn and winter, sometimes into spring. Seeds are usually only released from the woody cones when the branch dies or after a bushfire.
Variants: the habit is variable, from low and spreading to erect, while forms from the Queensland distribution area are more prone to produce multiple flower spikes.

CULTIVATION:
An adaptable species under cultivation, growing equally well in light or heavy soils provided adequate moisture is present. Acid soils are preferred and either a sunny or partially shaded position is satisfactory. Reasonably frost hardy once established, the author grows this species in an area experiencing winter temperatures as low as -5 °C. Rare early frosts have burnt soft young growth, however. A banksia which may be pruned severely to promote bushy new growth from the base or old stems.

PROPAGATION:
Always propagated from reasonably fresh seed usually sown in autumn.

USES:
An unusual and striking species for the home garden, *B. robur* is useful not only for its ornamental properties of both leaves and flower spikes but being so adaptable it can often be grown in parts of the garden where many other *Banksia* may not succeed. Not generally regarded as a suitable species for cut flowers, but the leaves dry well for dried arrangements.

PLATE 18: *Banksia robur.*

Zoë C

BANKSIA SERRATA

The Old Man Banksia of Australian fairy tales and commonly referred to as Saw Banksia or Red Honeysuckle, this species is one of the most well-known banksias of eastern Australia. Found growing naturally in Tasmania, Victoria, New South Wales and in southern coastal Queensland as far as Noosa, in sandy habitats, often close to the sea, sometimes binding the sand dunes from erosion. The Latin *serrata* translates to saw-edged, referring to the leaf margins.

DESCRIPTION:
Growth habit: varying from a robust shrub of only 1 to 2 m to a single-stemmed tree up to 16 m high, this species often matures to a gnarled and twisted tree with a thick trunk of up to 500 mm diameter and bearing many knotted branches of great character. The bark is furrowed in appearance and grey or brown.
Leaves: large, glossy green leaves up to 200 mm long and 40 mm wide with distinctively saw-toothed margins. The undersides may carry a grey down or can be glossy as with the upper surfaces.
Flowers: the cylindrical, terminal flower spikes measuring 70 to 150 mm in length and up to 120 mm in diameter when fully developed, are a beautiful, soft silvery-gold colouring. As the flowers age they assume a reddish toning before finally withering to grey. They are produced from mid summer to late autumn or even mid winter in some localities. Seed is released from the cases after bush fires or when the branch dies.
Variants: slight variations in flowers, leaves and growth habits occur throughout the range of this species, but it is quite uniform in its basic characteristics. New developments include dwarf and prostrate selections, which are proving popular in Australia.

CULTIVATION:
A slower-growing species than some, but generally long-lived in cultivation. Adaptable to both heavy and light soil types provided they are well drained. Either acid or slightly alkaline soils are suitable. *B. serrata* adapts well to both sunny or partially shaded positions and is reasonably frost tolerant. It may be pruned severely to encourage fresh growth and a more compact habit.

PROPAGATION:
As a rule this species is only propagated from seeds, which should be sown when still relatively fresh, usually in spring or autumn, into a free-draining medium such as peat and pumice or perlite or a sand/vermiculite type of mix.

USES:
Not a species generally grown for garden display, being too large for many plots, but it can make a handsome tree with character for background effect or shelter. The flowers are not considered to have commercial potential for picking. The seed cones are prominent and ornamental, however. The strong timber of this tree banksia has been used for various purposes, including boat building.

PLATE 19: *Banksia serrata.*

Zoë C.

BANKSIA SPECIOSA

Known by the appropriate common name of 'Showy Banksia', to which the Latin *speciosa* also translates. Grows naturally along the south coast of Western Australia, particularly in the Esperance area, in deep sand country.

DESCRIPTION:
Growth habit: a large, multi-branched shrub attaining a maximum height of 8 m with maturity, although rarely seen to this height in cultivation. Broadly branching in habit, reaching as much as 3 m in width, sometimes with stems reaching to the ground.
Leaves: particularly long, narrow and deeply serrated, the deep green leaves have a white tomentum on the undersides and reach to 450 mm in length and only 20 to 40 mm wide. They tend to hang downwards and are frequently quite undulating and wavy in appearance.
Flowers: the spectacular large flower spikes are silvery at first but then become a bright lemon-yellow or chartreuse shade. As the spikes develop they are first cylindrical, then broadly acorn-shaped until they are fully open when they are tightly drum-shaped. Carried terminally atop long, sturdy stems in summer and autumn, occasionally at other times of year, the spikes are spectacular when seen from a distance.
Variants: some variation in size of leaves and flower spikes is recorded but fundamentally this is a uniform species throughout its range.

CULTIVATION:
B. speciosa is a notoriously difficult species to grow in heavy soils, and gardens with deep, light soils that are sandy or gravelly usually produce the best results. The author has grown this banksia to perfection on a steep bank of light, sandstone-derived soil in an open, exposed situation. In this position it flowered continually for more than six months of the year and reached 2.5 m in height with an almost equal width within 4 years. Reputedly frost tolerant, *B. speciosa* can nevertheless have young growth damaged by early or late frosts. As a general rule it can withstand the occasional midwinter temperature drop to -5°C once established.

PROPAGATION:
Easily propagated from fresh seed which germinates quickly.

USES:
A species which is a useful cut flower, the blooms lasting well. It is not one of the commercial growers' first choices, however, due to the excessively long foliage and flower colour which is less popular than some. It is a popular choice with gardeners in coastal, sandy situations as it thrives in these conditions and is used as a large specimen shrub with picking qualities or for background effect.

PLATE 20: *Banksia speciosa.*

Zoë C.

BANKSIA SPINULOSA

Commonly known as Hairpin Banksia, for obvious reasons, and first described in 1792. There are three distinct varieties; *B. spinulosa* var. *collina*, *B. spinulosa* var. *cunninghamii* and *B. spinulosa* var. *spinulosa*. Located naturally throughout the east coast of Australia from Victoria, through New South Wales, to Cairns in Queensland. The Latin *spinulosus* means spiny, referring to the toothed foliage.

DESCRIPTION:

Growth habit: a bushy shrub to 3 or 4 m high, usually slow growing, attaining 1.5 m at 5 to 6 years from the seedling stage. It may also take as long to flower.

Leaves: the foliage is the main distinguishing feature between *B. spinulosa* var. *spinulosa* and var. *collina*, the latter variety having noticeably broader leaves with distinct veins. Var. *spinulosa* has almost needle-like foliage by comparison.

Flowers: Hairpin Banksia is an appropriate common name, the hooked styles giving this appearance. The elongated and cylindrical flowers may be up to 200 mm long and 40 to 80 mm in diameter. Colouring is basically honey or bronze, with the black or deep red hooked styles giving an overtoning of colour. They are produced freely on mature bushes over autumn and winter when they make a handsome display.

Variants: apart from the three distinct varieties, the length of the leaves varies with the locality and colour tonings of flowers may range from yellow to deep gold. A very dwarf selection, 'Birthday Candles', is available in Australia.

CULTIVATION:

Well-drained or heavier soils with adequate mositure and sun or semi-shade are all suitable for this species, serving to illustrate that it is adaptable to a wide range of conditions. It has been known to tolerate frost down to -8°C, but this may depend on seasonal conditions. Certainly it is a hardier species than many. As the Hairpin Banksia has a lignotuberous rootstock which enables it to spring away again after being burnt in bush fires, it can also be heavily pruned to turn an old woody bush into a vigorous, compact shrub. *B. spinulosa* var. *cunninghammii*, however, does not have a lignotuber and therefore should not be subjected to hard pruning.

PROPAGATION:

This is one of the few banksias which can be successfully propagated from semi-hardwood cuttings, usually in autumn. Hormonal treatment of cuttings is not necessary. Seeds are also widely used to increase this species.

USES:

A dense, bushy shrub, one of the easier species to grow, and standing hard pruning as it does has made this banksia popular for background effect and shelter and also for ornamental display in more conspicuous parts of the garden. It is not generally grown for cut-flower purposes.

PLATE 21: *Banksia spinulosa.*

Zoë C

DRYANDRA FORMOSA

Usually referred to as the Showy Dryandra. The Latin *formosa* translates to beautiful. Its natural habitat is southern Western Australia, particularly in the Stirling Range where it is found at all altitudes, even on rocky, windswept ledges.

DESCRIPTION:
Growth habit: a tall, dense shrub to 3.5 m, sometimes of more open habit when grown in hard conditions.
Leaves: the long, narrow leaves are deeply cut almost to the midrib, and have a saw-like appearance. Held erectly from the stems, particularly from young shoots, the foliage is held either horizontally or drooping downwards when it surrounds the flowers. Leaves are up to 120 mm in length and only 10 mm wide, the colouring is dark green, and greyish and densely hairy on the undersides.
Flowers: the terminal flowerheads are held conspicuously above the foliage and are frequently clustered in masses of both flowers and buds. Colour is variable from yellow to orange-bronze, but more usually is a golden bronze, the centre portion of the head a deeper bronze until it opens out. The flowerheads are produced over an extended period from late winter through spring into early summer, mid spring usually being the peak time. They may measure up to 60 or 70 mm across when fully developed, and 40 to 50 mm at the attractive, half-developed stage.
Variants: a fairly uniform species, *D. formosa* exhibits some variation in growth habit, and flowers may vary from yellow to orange-bronze shadings.

CULTIVATION:
As with most members of the protea family, this species requires a lighter, free-draining soil. Heavy soils inevitably seem to pose difficulties for successful cultivation. The author succeeded with this dryandra to perfection in a light, sandstone-derived soil where it reached 3 m in 4 years and flowered profusely. The only problem incurred was that root development seemed to be sacrificed in favour of the top growth and the bushes would often blow over in strong winds. This does seem a characteristic of this species, however. Light pruning encourages a more branched and compact habit.

PROPAGATION:
Normally propagated from the freely produced seed, but some limited success can be achieved from semi-hardwood cuttings. As the seed usually germinates readily and seedlings grow rapidly there would seem little purpose in vegetative propagation of this species.

USES:
A particularly ornamental, larger garden shrub, *D. formosa* also has some use for cut-flower purposes, the flowerheads lasting well and sometimes being dried.

PLATE 22: *Dryandra formosa.*

ZoëC

DRYANDRA POLYCEPHALA

Sometimes referred to as Many-headed Dryandra, which is also the translation of *polycephala*, also occasionally sharing the common name of Showy Dryandra with *D. formosa*. Found growing naturally in gravelly soils of the Darling Plateau of Western Australia.

DESCRIPTION:
Growth habit: an erect, often slender shrub growing to 3 m or sometimes 4 m in height.
Leaves: the narrow, prickly foliage is often curved in appearance, may measure to 80 mm in length and 10 to 20 mm wide. It is quite widely spaced on the branches thus making the flowers more conspicuous.
Flowers: regarded as the most free-flowering dryandra, this species produces massed displays of small, fluffy, bright yellow blooms, each measuring no more than 30 mm wide, from late winter through spring. They cluster quite densely along the long stems.
Variants: some natural variation occurs in flower colouring, some being of softer yellow tonings than others. Foliage is also variable, notably in the degree of 'prickliness', and growth habit is sometimes more branching and bushy as opposed to the more usual upright growth.

CULTIVATION:
Regarded as more frost tolerant than most dryandras, this species is relatively easy to grow under cultivation, but has the reputation of not being totally reliable in all soil types. Duplication of the well-drained gravelly soils of its habitat produce the best results, and gardens having this type of soil usually have good success. Some light pruning after flowering is desirable to produce a well-shaped bush.

PROPAGATION:
Usually propagated from fresh seed sown in autumn, this species can also be increased vegetatively from semi-hardwood cuttings.

USES:
A very ornamental garden shrub, *D. polycephala* is also used for cut flowers, the blooms lasting well and able to be dried.

DRYANDRA PATENS

A lesser known species. Occurs naturally in sands and gravels in south-west Western Australia. Not widely cultivated. *Patens* means spreading and refers to the floral leaves.

DESCRIPTION:
Growth habit: an erect, slender, bush or small tree, 3 to 5 m in height.
Leaves: deep green leaves, strongly toothed with sharp points, to 120 mm in length, and 20 mm wide, sometimes with a slight wave or twist.
Flowers: a light yellow colouring and rounded, similar to *D. polycephala*, although less clustered. Carried over late winter and spring. Not a cut-flower proposition.

CULTIVATION:
Similar requirements to *D. polycephala* (above).

PLATE 23: *Dryandra polycephala* (left), *Dryandra patens* (right).

DRYANDRA PRAEMORSA

Commonly known as Cut-leaf Dryandra or Urchin Dryandra. The Latin *praemorsa* refers to the blunt leaf tips and translates to 'bitten off'. In nature it is restricted to the south-western region of Western Australia, where it occurs mainly on well-drained granite soils.

DESCRIPTION:
Growth habit: a bushy shrub, attaining a height of 2 m or more with an equal spread, this species develops a distinct trunk with maturity and in old age can assume a slightly gnarled and small tree-like appearance.
Leaves: the foliage is a medium green shade, whitish on the undersides, broad (to 40 mm), up to 140 mm long and distinctively toothed with blunt ends. Noted for the resemblance to holly leaves.
Flowers: the showy, bright yellow, sometimes tinged pink, 'shaving brush'-like terminal flowers are carried in profusion on short stems during spring and early summer. They average 60 mm in both depth and width when fully open. Attractive in bud form also.
Variants: a variable dryandra, particularly in flower colouring, some forms being pure, golden yellow, others heavily flushed with a salmon-pink toning. Sometimes variable in growth habit also, from low and bushy to more upright.

CULTIVATION:
One of the easiest dryandras to cultivate in both Australia and New Zealand. Frost hardy in Canberra and many parts of New Zealand's South Island. Suitable for exposed situations provided some support in the form of a sturdy stake is given. Semi-shade in most Australian regions is preferable, but sunny positions in New Zealand appear suitable. Well-drained, acid soils are essential. Poorly drained soils may result in the early demise of shrubs due to root diseases. Light pruning after flowering is recommended to encourage a more compact, bushy growth habit. *D. praemorsa* does not seem prone to any particular insect problems under cultivation and is generally a long-lived shrub in most garden situations.

PROPAGATION:
Usually grown from seed, which may need to be treated using such methods as scalding with hot water or nicking to either soften or break the hard seed coating and initiate germination. May also be propagated from cuttings.

USES:
An attractive bush for many gardens, giving bright colour over several months, the unusual foliage adding interest when flowering has finished. It has some limited use as a cut flower for floral art purposes, but is not considered to have any commercial picking potential.

PLATE 24: *Dryandra praemorsa.*

Zoë C

DRYANDRA PROTEOIDES

Commonly known as King Dryandra. The specific Latin name refers to the protea-like flowers. An unusual and distinctive species rarely seen in cultivation. Endemic to Western Australia, where it is found growing on well-drained sandy or gravelly soils.

DESCRIPTION:
Growth habit: a bushy species, growing to 2 m by 2 m.
Leaves: foliage typical of many dryandra species is found on this shrub, 200 mm long, prickly and deeply cut and divided to the midrib, deep green but greyish-white on the undersides, often undulating and wavy in appearance.
Flowers: the most striking flowers (with the possible exception of *D. speciosa*) of any dryandra, being so typically protean in shape and form, with the prominent brownish-bronze bracts enclosing the cream flowers. No other species of this genus has such pronounced bracts. Measuring up to 100 mm across and 70 mm deep when fully developed, they are borne over late winter and spring.
Variants: usually a distinct and uniform species, only minor variations in growth habit and flowering time being noted.

CULTIVATION:
Not a difficult species to cultivate provided the typical requirements of very free-draining soil and a warm, sunny position are provided. Fertilisers, as usual, should be avoided. Light pruning, keeping within healthy foliage, will help to keep the bush compact.

PROPAGATION:
Normally propagated from seed, sown in the autumn or spring in a free-draining medium of peat and perlite or pumice, sand and vermiculite, etc.

USES:
Not a particularly popular garden subject, nor a cut-flower proposition, *D. proteoides* is more a connoisseur's item, being quite different from other species but very appealing in its own way.

PLATE 25: *Dryandra proteoides*.

Zoë C

GREVILLEA ALPINA

Commonly known as Mountain Grevillea, Cat's Claws or Goldfields' Grevillea. Despite its specific botanical name, which infers alpine origins, this species is rarely found at higher elevations, and more usually inhabits eucalypt forests through a wide distribution range — Victoria, New South Wales and Australian Capital Territory.

DESCRIPTION:
Growth habit: one of the most variable grevillea species, particularly in growth habit and flower colour. Its habit ranges from prostrate to a shrub of 2 m in height. More usually it is a small shrub of open growth habit to 60 or 70 cm tall.
Leaves: foliage is also variable, ranging from linear through oval to almost round. Typically, the leaves are oval, 20 mm long, slightly hairy, and a dull, grey-green colour.
Flowers: in the natural habitat this species can be seen bearing flowers of white, yellow, orange and red and frequently in combinations of these shades. The form known as *G. alpina* 'Dallachiana', one of the most commonly cultivated, has red and white flower clusters. These are borne from spring through summer into autumn.
Variants: as detailed in the previous paragraphs, this species probably shows more variation than any other of the genus. There are some 21 forms described in one publication, plus a registered cultivar, 'Grampians Gold', with golden yellow flowers and a lower growth habit. Ranging from prostrate to 2 m tall, with variable foliage and flower colours varying from white to red, *G. alpina* is a most confusing species for the amateur attempting identification in the field.

CULTIVATION:
Not regarded as a suitable species for humid areas, it is however very adaptable to dry, cool districts, tolerating both coastal and inland conditions and normally frost-resistant. It is also drought tolerant. Well-drained soils are essential. A light pruning after flowering, keeping within the foliage, assists in maintaining a compact, bushy habit.

PROPAGATION:
G. alpina is not normally an easy species to grow from seed, but cuttings usually strike readily. This may be as well as seedlings are very variable, and it is normally selected forms that are required commercially.

USES:
In larger rockeries, shrub beds and similar situations this smaller grevillea and its forms are at their best, the long flowering period making them very useful in a variety of landscapes. Generally regarded as one of the best small grevilleas.

PLATE 26: *Grevillea alpina.*

Zoë C

GREVILLEA BANKSII 'FORSTERS'

G. banksii is a particularly variable species and the selected variety 'Forsters' is the most widely cultivated form. The natural habit of this grevillea is south-eastern Queensland where it is found growing on stony hillsides and sandy ridges. Common names include Red-flowered Silky Oak, Red Silky Oak and Banks Grevillea. The specific *banksii* is in honour of Sir Joseph Banks.

DESCRIPTION:
Growth habit: while the widespread, naturally occurring form may produce an erect, large shrub or small tree to 8 m in height, var. 'Forsters' usually grows to 3 m by 3 m.
Leaves: deeply lobed, to 250 mm long, dark grey-green on the upper surface, silver on the underside with silken hairs.
Flowers: produced all year round, more especially in milder climates. They may measure 120 mm long, are borne terminally as cylindrical racemes in a striking combination of deep pink and bright red shadings and are attractive to nectar-seeking birds.
Variants: the exact origins of this variety are unknown but both it and the naturally occurring form are regarded as important for the qualities they impart to their progeny when used in hybridising. 'Robyn Gordon', 'Superb', 'Ned Kelly', 'Misty Pink' and 'Pink Surprise' are examples. Within the habitat varying forms include prostrate, fine-leaved and white-flowered. One of the latter has been given botanical status.

CULTIVATION:
Although easily grown in well-drained soils that are not too well enriched, this variety and the species are frost-tender and best suited to mild climates. Widely cultivated in Australia, New Zealand, South-East Asia and the warmer states of the United States. Regular pruning is essential to maintain a compact habit.

PROPAGATION:
Regarded as difficult to propagate from cuttings, seed is the more usual method as the resulting plants are normally true to type.

USES:
A popular shrub for the prolific production of the large, spectacular flowers. Used in mixed plantings and as a specimen.

PLATE 27: *Grevillea banksii* 'Forsters' (upper right), *Grevillea* 'Robyn Gordon' (lower right).

GREVILLEA 'ROBYN GORDON'

A deservedly popular and well-known hybrid originating at D.M. Gordon's property at Glenmorgan in Queensland and named for his late daughter. A chance hybrid between *G. banksii* and *G. bipinnatifida*, both of which are popular for hybridising purposes. The former species has Queensland as its natural habitat, while the latter is from Western Australia.

DESCRIPTION:
Growth habit: a broadly spreading, dense shrub of 1.5 m in height with a width of 2 to 3 m or occasionally more.
Leaves: the fern-like leaves are approximately 140 mm long, much divided and represent a truly hybrid mix between the parents, being much more divided and firmer than *G. banksii* and yet less divided and smoother than the other parent, *G. bipinnatifida*.
Flowers: the spectacular, large, pinkish-red flowers up to 160 mm long, droop in arches almost throughout the year, but with flushes in spring and autumn. Like the foliage, the flowerheads are intermediate between the parents, having the softness of *G. banksii* but the more intense colour and drooping habit of *G. bipinnatifida*. They are very attractive to nectar-seeking birds.
Variants: being a manipulated hybrid, natural variants do not occur. However, several other manipulated hybrids of the same cross are well known and popular. In Australia 'Ned Kelly' (also known as 'Mason's Hybrid') is popular for its more orange-red flowers, upright habit to 2 m, hardiness and disease-resistance. Less popular in Australia but widely grown in New Zealand is 'Superb', which has softer orange flowers and is intermediate between 'Ned Kelly' and 'Robyn Gordon' in growth habit. It is markedly more frost hardy than 'Robyn Gordon'.

Zoë C

CULTIVATION:
A relatively easy grevillea to grow, being adaptable to most reasonably well-drained soils. Not very frost tolerant, much more than light frosts causing bud damage. Warm, sheltered positions are preferred. The similar hybrids 'Ned Kelly' and 'Superb' tolerate frosts to -4 °C before damage begins to show up. Suited to either full sun or filtered shade. Fertilisers, particularly those containing phosphates, should be avoided. Regular, light pruning is advisable.

PROPAGATION:
As with all hybrid plants, vegetative propagation is essential. Not as easy as some, the timing being fairly crucial, as relatively soft material usually strikes more successfully. Some nurserymen find 'bottom heat' (the use of soil heating cables) helpful in encouraging root formation. Tissue culture is now the preferred method of propagation in Australia.

USES:
An outstanding grevillea for spectacular garden colour throughout the year. Excellent for attracting nectar-seeking birds. Not suitable for picking, the soft blooms usually drooping quickly when cut.

GREVILLEA 'BOONGALA SPINEBILL'

'Boongala' was the name of Sid Cadwell's New South Wales property where this cultivar originated; Spinebill refers to the honeyeater birds such as the Eastern Spinebill. A hybrid supposedly of *G. bipinnatifida* x *G. caleyi* origins. Although there are some doubts as to the authenticity of this, almost certainly *G. bipinnatifida* is one of the parents.

DESCRIPTION:
Growth habit: an attractive bushy shrub of 1.5 to 2 m in height with an often greater spread of 2 to 3 m. Tends to have a semi-pendulous or weeping growth habit with foliage to the ground.
Leaves: the relatively narrow (25 mm), very divided leaves have slightly prickly lobes, measure up to 120 mm in length, and are a deep, rather glossy green. Young growth is always tinged with a reddish-bronze colouring.
Flowers: the deep crimson-red 'toothbrush' flowers are borne prolifically all year. Up to 100 mm long, they are a feature of the bush and are attractive to some nectar-seeking birds. They tend to hang down like those of the parent *G. bipinnatifida*.
Variants: being a manipulated hybrid, natural variants do not occur. The cultivar *G.* 'Bronwyn Cleeves' is so similar that some believe it to be the same. An authenticated hybrid of *G. bipinnatifida* x *G. caleyi* exists which bears no resemblance to 'Boongala Spinebill', giving rise to the doubts regarding the latter's parentage.

CULTIVATION:
An especially adaptable grevillea, 'Boongala Spinebill' is noted for being tolerant of frost, drought and humidity. It is widely regarded as the hardiest grevillea for all landscaping purposes. The author has grown it in two gardens, the first in almost pure sandstone and a frost-free but windy position, the second in clay soil and a sheltered, frost-prone area; in both instances there was little or no variation in growth and vigour. Both situations provided sun and good drainage and were in soils that were low in nutrient levels. Those are probably the only prerequisites for successful cultivation.

PROPAGATION:
Always propagated vegetatively from semi-hardwood cuttings, usually taken in autumn, but can also be taken in late spring. Also successfully propagated by tissue culture.

USES:
An outstanding hybrid grevillea for a variety of landscaping situations. The semi-weeping habit is attractive, and the deep red flowers over the entire year plus the colourful young growth add further to its appeal. For sheer versatility in terms of adaptation to varying climates and soils it is probably out on its own. A good subject for screening purposes.

PLATE 28: *Grevillea* 'Boongala Spinebill'.

ZoëC

GREVILLEA DRUMMONDII

An unusual species, not often seen in cultivation but certainly a worthy garden subject. Occurs naturally in sandy loams in jarrah forest throughout a broad band of south-west Western Australia from near Perth to the Pemberton area. Commonly called Drummond's Grevillea and named for Charles Drummond (1784–1863), a noted plant collector.

DESCRIPTION:
Growth habit: variable, but generally a small shrub to 1 to 1.5m.
Leaves: ranging from broadly oval to almost round, and 30 to 50 mm in length, the leaves are one of the more intriguing features of this species. They are particularly soft and hairy, the quite long, silky hairs giving a halo effect to each leaf. Deep, grey-green in colour, this grevillea's foliage is quite unlike almost any other.
Flowers: also very distinctive, the small flowerheads measuring 30 to 40 mm across are an unusual colour, commencing as yellow at the bud stage, becoming orange, with red gradually assuming dominance as the flower develops. Some forms are almost bi-coloured distinctly orange and yellow, while others open from cream buds. The blooms are attractive to nectar-seeking birds such as the Spinebill and New Holland Honeyeaters in Western Australia.
Variants: apart from natural variations in flower colour which occur, there are three distinct named forms recognised as subspecies: *G. drummondii* ssp. *pimelioides*, the most commonly cultivated one; *G. drummondii* ssp. *drummondii*, the northern form, which has smaller leaves and cream flowers aging to red; and *G. drummondii* ssp. *centristigma* which is regarded as a miniature form with orange-yellow flowers. All three have distinct habitats and varying cultural requirements.

CULTIVATION:
Typical of most grevilleas, this species requires well-drained soils, ideally sandy loams or gravelly soils. However, the more usually cultivated ssp. *pimelioides* seems to prefer partially shaded positions and some summer moisture, while ssp. *drummondii* withstands drier, sunnier positions. The miniature form ssp. *centristigma* also requires more moisture. Apart from the above requirements, this species is adaptable to both humid and drier environments and is regarded as reasonably hardy. Some light pruning after flowering is beneficial in maintaining a neat, compact growth habit.

PROPAGATION:
Readily propagated from either fresh seed or semi-hardwood cuttings.

USES:
A delightful and unusual species for many landscapes. This one is not only quite showy for the flowers, but a focal point due to the attractive foliage, and has the added bonus of attracting nectar-seeking birds. *G. drummondii*, in all three forms, is increasingly sought after by grevillea connoisseurs.

PLATE 29: *Grevillea drummondii.*

Zoë C

GREVILLEA x GAUDICHAUDII

A natural hybrid between *G. acanthifolia* and *G. laurifolia* occurring in the Blue Mountains area of New South Wales. While *G. acanthifolia* is an upright shrub of 1 m or more, the other parent, *G. laurifolia* is a prostrate, dense, mat-forming plant, this habit having been passed on to the hybrid progeny that is generally cultivated. The name 'gaudichaudii', is for the Frenchman, Charles Gaudichaud-Beaupré (1789–1854), a noted botanist.

DESCRIPTION:
Growth habit: the cultivated form of *G.* x *gaudichaudii* is a much-branched, prostrate shrub reaching no more than 150 mm in height, but spreading out to form a carpet of foliage 4 m or more across.
Leaves: deeply cut, 80 to 100 mm in length, deep green, characteristically oak-leaf in shape but with a small red spike or 'prickle' at the tip of each of the 5 to 7 lobes. Young growth is a striking red colour.
Flowers: the 'toothbrush' flowers are a deep, wine-red colour, 40 to 60 mm long, and are present over late winter, spring and summer. They are carried particularly prolifically when the plant is in a warm, sunny position, but as with most of the Proteaceae, this grevillea is shy to flower in shaded places.

CULTIVATION:
Most well-drained soils ranging from sandy to clay types are suitable. Sun is a prerequisite to a good flower display. Reasonably frost tolerant, although young growth can be burnt by early frosts. It is resistant to salt spray, making it a good coastal garden subject.

PROPAGATION:
Always propagated from cuttings of semi-hardwood material.

USES:
A superb ground cover plant. Has been used for grafting onto standards, but not as effective as *G.* 'Royal Mantle'. Foliage sometimes used by floral artists.

GREVILLEA 'POORINDA ROYAL MANTLE'

A hybrid of *G. laurifolia* and *G. willisii*. First registered in 1975 by well-known grevillea hybridist Leo Hodge. Many of his hybrids are prefaced 'Poorinda', after the name of his property in Victoria.

DESCRIPTION
Growth habit: a similar growth habit to that of *G.*x *gaudichaudii* but even flatter and more prostrate and vigorous, easily spreading to over 6 m in width.
Leaves: variable, some leaves being entire, others distinctly lobed, the latter being considerably larger, measuring to 100 mm long and 40 mm wide, while the entire ones are usually no more than 90 mm long and half as wide at 20 mm. A medium green colour; the young growth, however, is covered in fine, silken hairs and is an attractive bronze shading.
Flowers: a light wine-red colour, 'toothbrush'-shaped, 50 to 60 mm long and borne terminally on the short branchlets and sometimes on the main stems, they are prominently displayed over a long period during spring, summer and through autumn into winter. There is only a short period when no flowers are evident.

CULTIVATION:
Regarded as the hardiest of all ground cover grevilleas, *G.* 'Poorinda Royal Mantle' is notably frost hardy, drought resistant and even tolerates humid, tropical climates. It is adaptable to most well-drained soils and flowers more profusely in sunny positions.

PROPAGATION:
Always propagated from cuttings of semi-hardwood material, usually in autumn.

USES:
An outstanding, tough ground cover plant, covering large areas in a short time. Also used for grafting onto standards of *G. robusta* to create handsome weeping specimens.

PLATE 30: *Grevillea* x *gaudichaudii* (top right), *Grevillea* 'Poorinda Royal Mantle' (lower left).

Zoë C

GREVILLEA GLABRATA

A Western Australian species found growing in woodland conditions, often in damp situations. Not cultivated widely in Australia, but deservedly popular in New Zealand. *Glabrata* is from the Latin *glabratus*, meaning nearly glabrous or smooth.

DESCRIPTION:
Growth habit: a vigorous, quick-growing shrub to 3 m high with a greater width. Has an attractive weeping habit.
Leaves: slender, light green leaves are distinctively tri-lobed, each lobe ending in a point. They are 30 to 50 mm long and being borne on the pendent branches create an overall effect of gracefulness that makes this species popular as a foliage plant.
Flowers: small, delicate clusters of white flowers are borne at any time of year, but especially through winter and spring into early summer. Set against the fine, weeping foliage they add to the graceful appearance. A slight, pleasant scent is sometimes apparent.
Variants: some variation occurs in size and foliage, three distinct subspecies being recorded; ssp. *glabrata* being the best known. The other two are ssp. *dissectifolia* with more deeply divided foliage and ssp. *ornithopoda* with wedge-shaped leaves and taller growth.

CULTIVATION:
A relatively easy species under cultivation, but regarded as borderline for humid climates. Most well-drained soils suit, and either full sun or partial shade. It is recognised as being a hardy grevillea, tolerating quite hard frosts once established. Pruning can keep it to a smaller, compact shape.

PROPAGATION:
Readily propagated from semi-hardwood cuttings which should be well-rooted before pricking out into containers.

USES:
As already mentioned, *G. glabrata* finds ready favour as a foliage plant. It is also used as a quick-growing screen plant and the branches are sometimes cut for use as foliage in floral arrangements. It can be used to good effect as a solitary specimen and is sometimes pruned and trained to make a broad, weeping standard.

PLATE 31: *Grevillea glabrata* (top), *Grevillea curviloba* (bottom).

GREVILLEA CURVILOBA

Sometimes known as *G. tridentifera* 'prostrate', and, more commonly, as *G. biternata*, this species is found growing naturally in both South Australia and Western Australia, although the latter state is believed to be the origin of the form generally cultivated. Occurs naturally inland and on mountains.

DESCRIPTION:
Growth habit: a mat-forming prostrate species, particularly in its juvenile stages. It tends to throw odd, tall, erect branches, however, and these need to be removed to retain the prostrate habit. Left to their own devices these branches eventually transform the plant into a dense shrub of 1.5 m
Leaves: an easily recognisable and identifiable feature of this species, the leaves are light green and very finely divided to 50 mm in length.
Flowers: the white flowers in dense clusters are lightly tinged green, perfumed with a honey-like fragrance and appear for a prolonged period over spring.
Variants: some forms are more prostrate than others and are therefore sought for propagating from.

CULTIVATION:
This species can be difficult to grow successfully in humid climates and heavy soil types and is regarded as moderately frost tolerant. The dense growth can lead to fungal problems and some removal of older, underlying foliage may be necessary to prevent this. Light sandy soils

Zoë C

are among the best for successful cultivation of this species. The author has grown this grevillea to perfection on a bank of poor, sandstone-derived soil in a hot, dry situation and encountered none of the problems peculiar to the species. In heavier soil in a colder climate, on the other hand, all the problems have been encountered.

PROPAGATION:
Readily propagated from semi-hardwood cuttings which should be well-rooted before pricking out into containers.

USES:
Where soil and climate are suitable, *G. curviloba* is a superior ground cover, rapidly covering an area of 3 m square in 2 to 3 years. In such conditions, the dense, mat-forming, light green foliage is most attractive and the white flowers are set off against this to perfection. Upright growth will need to be removed to maintain the low habit.

GREVILLEA HOOKERIANA — HYBRID FORM

A confusing grevillea from several points of view. The spelling is in dispute, varying between *'hookerana'* and *'hookeriana'* and the origins of the form in cultivation are quite obscure. The species, found growing near Ravensthorpe in Western Australia, is more erect in habit and much less vigorous. The form commonly in cultivation is widely referred to in Australia as *G. hookeriana* 'hybrid', and in New Zealand is sold simply as *G.* 'Robin Hood'.

DESCRIPTION:
Growth habit: a particularly fast-growing, vigorous, dense, bushy shrub quickly reaching a height of 3 m with a nearly equal spread. In 2 years it is usually up to 2 m or more and it flowers almost immediately; within months of being propagated.
Leaves: characteristically 'fishbone' in appearance, being deeply divided, dark glossy green and up to 150 mm long. They are slightly prickly, each of the approximately 20 lobes ending in fine points. The undersides of the leaves are a silver-grey colour and display the midrib prominently.
Flowers: the rich, crimson-red, 'toothbrush' flowers are produced in abundance over most of the year, but are especially prominent over spring and summer. Measuring up to 100 mm in length but averaging 70 to 80 mm, they are attractive to birds and being produced so freely throughout the plant create a colourful display.
Variants: varying forms of the species range from the more compact to others with orange, brown and even black flowers.

CULTIVATION:
Another of the 'easy' grevilleas, *G. hookeriana* 'hybrid' is tolerant of heavier soils although still preferring well-drained, lighter conditions. It will stand up to more humid climates well and is frost tolerant to about -10 °C in most areas. Younger plants will stand quite hard cutting back, while older specimens will need more careful pruning, keeping cuts within the foliage area of branches. Fertilisers should be avoided, particularly those containing phosphates. This grevillea thrives in poorer soils. Not a good subject for windy, exposed situations, where branches tend to break off.

PROPAGATION:
A reasonably easy subject to strike from semi-hardwood cuttings.

USES:
An excellent, fast-growing screen plant, also suitable for specimen use, shrub borders and other landscape positions. The very free-flowering habit, virtually throughout the year, is just one of the appealing features. Attractive to nectar-seeking birds. One of the few grevilleas used for cut-flower purposes, the foliage is also picked, although this is usually restricted to winter when there is less soft young growth present which can droop when picked.

PLATE 32: *Grevillea hookeriana* 'hybrid'.

GREVILLEA JUNIPERINA

A species, in all its varying forms, restricted to New South Wales where it occurs both on the coast and the tablelands, usually growing in clay soils. The specific name refers to the juniper-like leaves.

DESCRIPTION:
Growth habit: variable, but generally a rounded or spreading shrub to 2 m in height with a spread of up to 2.5 m.
Leaves: narrow, needle-like and sharply pointed, medium green leaves are characteristic of this species and vary in length from 10 to 30 mm. There is a more broad-leaved form.
Flowers: amazingly variable in colour, ranging from yellow to red with all the intermediate shades represented, the definite reds and yellows being the most usual. Carried as dense clusters, 30 to 40 mm in width and appearing at various times, mainly winter, spring and early summer.
Variants: as already mentioned, this is a very variable grevillea. There are prostrate forms, erect forms and bushy examples. Colour of flowers and size of leaf also varies. It has been the basis for extensive hybridising and there are many named cultivars registered — too many to list in this text.

CULTIVATION:
Widely regarded as one of the hardiest and most adaptable grevilleas. Suited to growing in almost any well-drained soil, even heavier clays, in a sunny position. Stands quite hard frosts in most regions, examples surviving in districts experiencing occasional midwinter temperature drops as low as -20°C. Tolerates poor soils and fertilisers should be avoided.

PROPAGATION:
Easily grown from seed, but the resulting progeny can be very variable. Selected forms should always be grown from cuttings. These should be taken from semi-hardwood material in late spring or autumn.

USES:
Widely used in a variety of landscaping situations, particularly due to its easy-care requirements, hardiness and adaptability. Usually a very colourful, free-flowering shrub, *G. juniperina* is probably one of the most commonly planted species in all its varying forms in Australia, New Zealand, and even in California and the United Kingdom. Many nectar-seeking birds are attracted to the flowers.

PLATE 33: *Grevillea juniperina*, colour variants.

Zoë C

GREVILLEA 'McDONALD PARK'

More correctly known by the full title of *G.* 'Austraflora McDonald Park', this hybrid is of *G. rosmarinifolia* x *G. alpina* origins and was marketed by the well-known Victorian native plant nursery, Austraflora. It is a registered cultivar but in New Zealand it is also sold as 'Highland Laddie'. Both parent plants have a similar distribution range, having Victoria, New South Wales and the Capital Territory in common, while *G. rosmarinifolia* also occurs in South Australia. McDonald Park in Ararat, Victoria, is where this cultivar was found growing in 1967.

DESCRIPTION:
Growth habit: generally a low-growing, dense, spreading shrub, which according to references grows to only 20 cm high and 1 m wide. The author has found, however, that this cultivar more usually reaches to 70 cm in height with a spread of 1.5 m, although on a steep bank or planted on the edge of retaining walls it can develop an almost prostrate or cascading habit.
Leaves: narrow, entire leaves are a light to medium green and measure up to 30 mm long by only 3 mm wide. The dense foliage helps to create a compact appearance.
Flowers: 30 to 40 mm wide-clusters of typically 'spider' flowers are borne prolifically throughout winter with odd blooms appearing in spring and summer. The relatively small size of the flowers is more than made up for by the vivid colouring; virtually a bi-colour arrangement of bright scarlet and yellow, giving an overall rich orange appearance.
Variants: there may be similar hybrids occurring naturally.

CULTIVATION:
One of the easier grevilleas to cultivate, *G.* 'McDonald Park' has typically hybrid vigour, reaching its maximum size with dense growth within 2 to 3 years in most situations. It is adaptable to most average, well-drained soils and performs very satisfactorily, particularly from the flowering point of view, when planted in sunny positions. The author has always found this plant trouble-free and quite frost tolerant once established. It does not grow well in warm, humid climates. An occasional light pruning in spring will help to keep bushes compact.

PROPAGATION:
As with all cultivars, this grevillea has to be propagated from cuttings to ensure it stays true to form. Semi-hardwood material taken in late spring or autumn is usually most satisfactory. It is not difficult to strike.

USES:
A particularly useful grevillea for many landscaping situations. Able to be pruned and trained to form either a neat, rounded bush, or a low, spreading semi-ground cover subject. The vivid colouring of the flowers over the duller winter months is especially valuable.

PLATE 34: *Grevillea* 'McDonald Park'.

GREVILLEA ROBUSTA

Commonly known as Silky Oak, not only in its natural habitat, but almost wherever it is cultivated in many parts of the world. This name refers to the silky textured, oak-like timber. The Latin *robusta* refers to the large, strong growth habit. Its natural habitat and range is in the rainforests of northern New South Wales and southern Queensland.

DESCRIPTION:
Growth habit: a tall, straight tree of 30 to 40 m in height in its natural state, but more usually assuming a pyramidal shape in cultivation with a height rarely exceeding 16 m.
Leaves: the very ornamental and graceful fern-like foliage consists of pinnate leaves with lobed segments. Colouring is a mid-green with a glossy sheen. In cooler climates *G. robusta* may almost become semi-deciduous with a considerable number of older leaves dropping during winter. Spring sees regrowth of new foliage.
Flowers: spectacular flowers are a feature of this species with a mature tree in full bloom a magnificent sight. They are a rich golden or orange-yellow shade, carried as long, brush-like racemes of up to 150 mm during spring or early summer.
Variants: a uniform species with little or no variation apart from one specimen that has been recorded with golden foliage.

CULTIVATION:
An adaptable grevillea, taking either well-drained or heavier soils provided adequate moisture is present. Full sun is necessary to promote free flowering. Light frosts are unlikely to affect young plants and larger, well-established specimens appear to be able to tolerate winter temperature drops of up to -7 °C.

PROPAGATION:
Normally grown from seed, which germinates quickly and easily. Can be grown from cuttings if a specific form is to be perpetuated.

USES:
A grevillea with many uses, *G. robusta* is well known in Australia for its silky, attractive timber which has a multi-directional grain useful for such diverse purposes as boomerangs, cabinet work and plywood. It is also used as an understock for grafting; 2 or 3 m standards being used on which to graft different prostrate hybrids and species to create very successful weeping standards. The flowers attract both birds and bees.

Its more obvious and well-known uses are as a spectacular specimen tree for the larger garden or for an indoor tub plant while young.

PLATE 35: *Grevillea robusta.*

Zoë C

GREVILLEA ROSMARINIFOLIA

Occurs mainly in New South Wales and Victoria in both sandy and rocky terrain. The specific name *rosmarinifolia* refers to the rosemary-like foliage and it is commonly known as 'Rosemary Grevillea'.

DESCRIPTION:
Growth habit: a very variable species, ranging from prostrate to large shrubs of 3 m by 3 m. The majority of cultivated forms are in the range of 1.5 to 2 m and are densely bushy in habit. (See 'Variants' below.)
Leaves: foliage is typically narrow and pointed, needle-like, up to 35 mm long, and usually dark green. One form has smaller, grey leaves.
Flowers: as variable as the growth habit and leaves, the flowers are always similar in basic shape, being dense, rounded and 'spider-like', up to 50 mm in diameter, and are borne throughout the year with flushes in spring and autumn. However, colour may vary from the most common pink and white to deep red, cream and some with lime green and yellow flowerheads.
Variants: noted for being a very variable species, ranging from almost prostate to tall bushes of 3 m, with intermediate forms such as 'Pink Pixie', with light pink flowers and growing to only 500 mm. The grevillea known as 'Jenkinsii' is now thought to be a form. It has attractive deep pink and white flowers. There are countless hybrids of *G. rosmarinifolia*, even examples with variegated foliage as is the case with 'Red Dragon' a hybrid with 'Crosbie Morrison', itself a hybrid between *G. rosmarinifolia* and *G. lanigera*. This is certainly a much favoured species to use as a basis for hybridising, its natural vigour, hardiness and adaptability giving rise to very reliable progeny, an important consideration in today's demanding marketplace.

CULTIVATION:
Regarded as one of the hardiest and easiest grevilleas, this species nevertheless has certain preferences and is recognised as not being tolerant of very humid climates. It will tolerate frost to -7 °C or more, can be grown equally successfully in coastal or inland situations, and will adapt to almost any well-drained light to medium soil type. It will withstand hard pruning, so much so it is sometimes used as a hedging subject.

PROPAGATION:
Easily grown from either seed or cuttings. Selected forms should only be propagated from cuttings. Seedlings usually prove quite variable.

USES:
A quick-growing, hardy and reliable larger shrub for using as a screen plant, for background or medium height shelter. Can be pruned hard and is used as a hedge. The flowers are also noted for attracting birds. Not a particularly spectacular species to use if a showy specimen is required.

PLATE 36: *Grevillea rosmarinifolia*.

GREVILLEA SPECIOSA

Previously known as *G. punicea*, it occurs exclusively in New South Wales in the Sydney region where it is found growing mainly on soils of sandstone and gravel. Commonly known as Red Spider Flower. The specific name *speciosa* means 'showy' or 'beautiful'.

DESCRIPTION:
Growth habit: there are many growth forms of this species, ranging from low, spreading shrubs of only 500 mm to 1 m high with a similar width, to erect examples of 3.5 m high with a spread of little more than 1 m. The form more usually seen growing in cultivation is a dense, bushy shrub of 2 to 3 m.
Leaves: once again these are variable, but the more commonly grown form has oval, pointed leaves that are dark green and 20 to 30 mm long. Other forms have light green leaves which can be quite narrowly lanceolate in shape.
Flowers: a bright red colour, the conspicuous flowerheads can measure 50 to 60 mm across and look like a wire-spoked wheel with a large central hub. The main flowering period is over winter to spring, but blooms appear at other times of the year as well. The best forms cultivated produce an abundance of flowers, while others bear fewer flowers of smaller dimensions.
Variants: a very variable species, as already discussed in the preceding paragraphs. While growth habit, leaf shape and colour, and flower size can all vary substantially, flower colouring remains constant — always a rich, bright red. There are two recognised subspecies.

CULTIVATION:
Best specimens in cultivation are usually seen growing in well-drained soils in sheltered, sunny positions. Frost tolerant to -5 °C, possibly more in some areas. Responds well to occasional light pruning and this maintains a compact growth habit. Fertilisers should be avoided, particularly those containing phosphates. Regarded as a very reliable garden plant.

PROPAGATION:
While this species is readily propagated from either seed or cuttings, seedling plants often prove variable. Cutting grown plants stay true to the parent characteristics which is why good, selected forms should always be propagated in this way.

USES:
A fast-growing, very ornamental grevillea well suited to a variety of landscaping situations. As a specimen, in shrub borders or for background effect, this is a lovely species. Widely recognised as probably the best of the red-flowered 'spider' type of grevillea, it is surprising that *G. speciosa* is not more widely used by gardeners. Attractive to some nectar-seeking birds.

PLATE 37: *Grevillea speciosa*.

Zoë C

GREVILLEA THELEMANNIANA

Occurring naturally over a broad range in Western Australia between Geraldton and south of Perth, the various subspecies tending to have separate distribution areas within that range.

DESCRIPTION:
Growth habit: variable habits characterise this species and its subspecies, ranging from dense, rounded shrubs of 1 m or less to the better known procumbent, ground covering forms which can spread to 3 m wide while never attaining a height of more than a few centimetres.
Leaves: the 20 to 40 mm long, slightly prickly and divided leaves vary in colour from a soft, woolly grey in the most prostrate form to bright green in another slightly more upright growing form.
Flowers: deep, scarlet red, 'spider' flowers in clustered heads up to 40 mm in length are carried profusely over winter and spring with blooms appearing into summer. Other colour forms include orange and pink.
Variants: there are many variations. There are at least seven recognised subspecies, with greatly varying foliage, some hairy, some smooth, some grey-green, others bright green. Growth habit varies from 1 m bushes to totally prostrate forms and flower colour ranges from red to orange and pink shades. Most are recognised as having good garden potential, but generally only two of the prostrate forms are readily obtainable by gardeners.

CULTIVATION:
Most light to medium, average, well-drained soils suit this species and its variants. Either full sun or partial shade is suitable, and while it is not recognised as being particularly frost tolerant, the author has seen and grown the grey-foliaged prostrate form in areas experiencing midwinter frosts of -5°C, where it thrived to perfection. This latter form is ideally suited to cascading over quite high walls and specimens have been noted spilling down retaining walls to a fall of 3 m or more and covering a width of over 2 m. It is susceptible to damage from fertilisers and these should be avoided. Prune lightly to encourage a dense, branching habit.

PROPAGATION:
Usually propagated from cuttings of semi-hardwood material taken in late spring or autumn.

USES:
The prostrate, grey-foliaged form, which is one of the most commonly grown and is the version illustrated, is best grown in a position where it can hang over a wall or sprawl down a steep bank. In such situations it can be seen to best advantage and is usually more easily maintained. If grown on flat ground, falling leaves from other plants are hard to extract from the foliage. In the right situation, a superb grevillea.

PLATE 38: *Grevillea thelemanniana.*

Zoë C

GREVILLEA VICTORIAE

Commonly known as Royal Grevillea or Mountain Grevillea. Occurring naturally in high altitude areas of Victoria, New South Wales and Australian Capital Territory where it is mainly found growing in rocky soils. Named for Queen Victoria.

DESCRIPTION:
Growth habit: very variable in nature from low or straggling, to compact and rounded or taller and more erect. Generally, particularly with the cultivated forms, a dense, rounded shrub of some 2 to 2.5 m in height with a spread of 1.5 to 2 m.
Leaves: unusual among grevilleas, the narrow, entire leaves are a distinctive dark green colour, grey on the undersides, and measure up to 100 mm in length. Overall appearance of the foliage is a silvery green. One form from near Canberra has almost round leaves to 60 mm. Young growth has a bronze tint.
Flowers: the deep, rusty red buds are a feature of this plant, hanging in prominent clusters, and having an almost furry appearance. They open to form ornamental, pendent clusters of dull, brick-red flowers 50 to 80 mm long. Flowering time is variable and extended, mainly from winter through spring, and blooms are attractive to birds. One form has pink flowers.
Variants: as already mentioned, this is a very variable species. Forms are recorded with prostrate habits, one with a pendulous habit and pink flowers from Lake Mt. Victoria, some with round foliage, some more compact in growth habit, to name just a few. Two subspecies are recognised, var. *victoriae* and var. *leptoneura* which has smaller leaves and flowers, a more open habit and grows in moister, shadier positions at lower altitudes.

CULTIVATION:
Coming as it does from higher, colder altitudes, *G. victoriae* is more frost tolerant than many grevilleas, and provided it is growing in well-drained situations should withstand frosts of at least -7 °C. Once established it is drought tolerant also, and will survive under large trees where many other plants could not. On the other hand, it will also usually withstand quite heavy, moist soils and partially shaded situations. Best results are obtained in well-drained, average soils in sunny positions.

PROPAGATION:
Because it is variable when grown from seed and is so readily propagated from semi-hardwood cuttings this is generally the preferred method.

USES:
A valuable species for use as a specimen, for a fairly quick-growing screen plant or to attract nectar-seeking birds. The prolific flowering habit gives colour in the garden for months at a time.

PLATE 39: *Grevillea victoriae.*

HAKEA LAURINA

Commonly known by various names, including Pincushion Hakea, and Pincushion Plant. Has a limited natural distribution in parts of the south coast and inland sand plains of Western Australia. The genus *Hakea* is restricted to Australia and there are more than 100 species, *H. laurina* being one of the best known and most spectacular. *Hakea* is named for a German patron of botany, Baron von Hake and *laurina* for the plant's laurel-like foliage.

DESCRIPTION:
Growth habit: in its natural habitat this species is usually seen as a slender shrub or small tree, rarely more than 2.5 m in height. Under cultivation, however, it can quite rapidly reach 5 m or more in height and assume a dense, rounded habit, or, with maturity, become a tree with a broad, multi-branched canopy of foliage.
Leaves: often likened to those of many eucalyptus species, the leaves are elliptical, deep green, to 150 mm in length and prominently longitudinally veined. Young growth is a soft, golden-bronze, while old foliage frequently becomes orange or red before falling.
Flowers: the globular shaped 'pincushion' flowers are golf-ball-sized, and basically red with large numbers of protruding cream styles like pins. The overall colour effect is frequently more of an orange shade. They open from unusual buds that are initially covered in brown bracts. These peel off to reveal the unopened flower in pale cream, which then rapidly opens. Blooms are borne prolifically over autumn and early winter and are attractive to nectar-seeking birds.
Variants: no significant variants are recorded, apart from minor variations in growth habit.

CULTIVATION:
An adaptable species under cultivation, *H. laurina* is tolerant of both light and heavy soils, provided they are well drained and not excessively rich. It will tolerate some frost once established, but young growth can be damaged by early frosts. A species that is rather notorious for being shallow-rooted and easily uprooted by strong wind, *H. laurina* should be planted in a sheltered position in full sun. Cooler climates can be detrimental to good flowering, it often being said that hot, dry summers are necessary to 'ripen the wood'.

PROPAGATION:
Normally propagated from seed, which should be sown fresh for best germination.

USES:
A popular large shrub to small tree for spectacular autumn flower display. Probably best suited to larger gardens. The author has seen this species used successfully as a hedge where regular clipping kept it very dense. Not regarded as a cut-flower species, although branches of bloom are used to spectacular effect by floral artists.

PLATE 40: *Hakea laurina*.

ISOPOGON ANEMONIFOLIUS

One of the approximately 34 species of this genus which mainly occurs in Western Australia, *I. anemonifolius* is found in the eastern states of Victoria, New South Wales and Queensland. Like most of the species, this one is also commonly known as Drumsticks. The genus was first described by the botanist Joseph Knight in 1810. *Anemonifolius* refers to the anemone-like foliage.

DESCRIPTION:
Growth habit: usually an upright shrub to a maximum of 2 m with a spread of 1.5 m, this species can occasionally be seen as a lower, spreading plant with a spread equalling its height.
Leaves: the light green foliage, much divided and stiffer than species such as *I. anethifolius*, is up to 100 mm in length. Young growth is an attractive, pinkish-bronze shading.
Flowers: the prominent, terminal heads of bloom are a rich yellow in colour and when fully open may measure up to 50 mm across. Carried on the bush over spring and early summer, they are always a feature. The characteristic 'drumstick' seed cones follow.
Variants: low-growing, almost prostrate forms, have been observed in the natural habitat and these may find their way into cultivation.

CULTIVATION:
An adaptable and relatively easy species to cultivate, *I. anemonifolius* needs well-drained soil and an open, sunny position to perform well. Regarded as hardy in Australia, the author has grown this species in an area where winter temperatures occasionally dropped to −4 °C. Light pruning after flowering, keeping cuts within foliage, is desirable to encourage a compact, bushy habit. Generally resistant to salt-laden winds.

PROPAGATION:
Usually propagated from fresh seed sown in autumn, but where this is unavailable semi-hardwood cuttings may be taken which normally root easily using standard techniques.

USES:
An attractive, ornamental shrub for the home garden, the flowers giving a good, colourful display over 2 months or more. Useful for floral art purposes also.

PLATE 41: *Isopogon anemonifolius.*

Zoë C

ISOPOGON ANETHIFOLIUS

The approximately 34 species of this genus are collectively known as 'Drumsticks'. They are found only in Australia where they are mainly restricted to Western Australia, with some occurring in Victoria, New South Wales and Queensland. *I. anethifolius* has New South Wales as its natural habitat. *Anethifolius* refers to the dill (anethum)-like leaves.

DESCRIPTION:
Growth habit: an erect, densely bushy shrub to 2 m, with a spread of 1.5 m, this species can become woody and more open with age.
Leaves: the attractive foliage is finely divided, medium green, with young growth and stems often tinged red. Individual leaves may measure to 120 mm or more in length.
Flowers: the prominent flower cones are carried on the ends of the branches. Opening from tight grey 'drumstick' buds to become round, bright yellow flowerheads, they are followed by large, globular heads of seed cones 25 mm in diameter which also resemble drumsticks — hence the common name. The main flowering season is spring and early summer.
Variants: some variation in growth habit has been observed, but this species is generally true to type throughout its distribution range.

CULTIVATION:
I. anethifolius, as with most of the eastern Australian species, is easier to grow under cultivation than the more temperamental Western Australian examples. Most well-drained soils and sunny positions suit, and this species is tolerant of some frost, the author having cultivated it successfully in a garden receiving midwinter temperatures as low as −4 °C on occasion. Early frosts in autumn can damage any young growth present, particularly on young plants.

PROPAGATION:
As with most members of the genus, this species is easily grown from fresh seed, which, if available, is the preferred method, quickly producing bushy, young plants of vigorous habit. Propagation from cuttings is not considered difficult either, but resulting plants often have a more pronounced trunk, although they do have the advantage of flowering sooner.

USES:
A highly ornamental shrub, this species has the dual attractions of a very showy flower display over a prolonged period plus attractive foliage at all times, but especially when the young red-tinged growth appears. It can be grown as a specimen or among other shrubs in borders. Flowers and cones are popular with floral artists.

PLATE 42: *Isopogon anethifolius.*

Zoë C

ISOPOGON DAWSONII

Another of the Drumstick type of isopogon, this comparatively little-known species is found in New South Wales, usually in lighter, well-drained soil types. Named for a New South Wales surveyor, Dawson, who discovered this species.

DESCRIPTION:
Growth habit: under cultivation generally a neat, erect growing shrub to a height of little more than 1.5 m, with a spread of less than 1 m. In nature it may reach 6 m as it is drawn up to the light through thick undergrowth.
Leaves: deep green foliage is divided and re-divided into slender segments to give an almost fern-like appearance. Each leaf may reach a maximum length of 100 mm. The young growth has a delicate bronze tinge giving a particularly pleasant overall effect to the bush.
Flowers: the creamy-white flowerheads measure 40 to 50 mm across when fully open and are highlighted by the protruding, yellow stamens. They are produced on quite long stems, usually over late winter and the spring months, for an extended period. The flowers are followed by small, typically 'drumstick' seed cones which persist on the bush for months.
Variants: no variations to the typical form are recorded.

CULTIVATION:
As is typical of the eastern isopogons, this is one of the easier species to grow in the home garden. Most average, well-drained soils and open, sunny positions appear to favour successful cultivation. A reasonably frost-tolerant species, the author having grown an example for many years on the lower slopes of a hillside where frosts of –3 °C were not uncommon in mid winter. As with most members of the family Proteaceae, fertilisers should be avoided. An occasional light pruning after flowering, keeping cuts within foliage, assists in keeping the growth bushy and compact.

PROPAGATION:
Not a difficult species to propagate, seed being the preferred method, but cuttings of semi-hardwood taken in autumn are also used.

USES:
An attractive garden plant, the soft tonings of the flowers blending well with others. The foliage is also appealing, adding further to the versatility of this species. As well, the reasonable stem length and good lasting qualities of the cut blooms make them useful for both florists and floral artists.

PLATE 43: *Isopogon dawsonii.*

ZoëC

ISOPOGON LATIFOLIUS

Possibly the most spectacular of the 25 Western Australian species of isopogon, *I. latifolius* is seen at its best growing on the slopes and high ledges of the taller peaks of the Stirling Range in south Western Australia. The Latin *latifolius* refers to the broad leaves.

DESCRIPTION:
Growth habit: a large, bushy shrub of 2 to 3 m in ultimate height with a spread of 1.5 m or more. It is rare, however, to see specimens as large as this in cultivation, the author's original plant reaching 1.5 m by 1.5 m after 5 years.
Leaves: this species shares with its close relative *I. cuneatus* the broadest leaves of the genus. Entire, leathery, faintly margined with a red edge, the rich green leaves may measure 100 mm in length and 40 mm in width. They are a feature of the bush and provide a handsome foil to the spectacular blooms.
Flowers: large terminal heads of rich mauve-pink flowers up to 100 mm across are carried on strong stems during spring. In nature, specimens growing at higher altitudes (to 1000 m) flower up to a month later than those at lower levels.
Variants: no conspicuously different variations are recorded.

CULTIVATION:
As is frequently the case with the Western Australia species of isopogon, this one is more difficult to grow under cultivation than the eastern examples. Light, well-drained soils and sunny, open positions are recommended. In Australia a 'limestone underlay' technique is recommended, consisting of limestone chips in a 100 to 150 mm layer with an acid, sandy soil type (to 300 mm deep) on top of it in which the shrubs are planted. *I. latifolius* is not a very frost-tolerant species and warmer positions in the garden will need to be selected.

PROPAGATION:
Usually propagated from fresh seed, but some success with cuttings is possible.

USES:
Where it can be grown successfully, this most spectacular of isopogons is a beautiful specimen shrub for the garden. In full bloom, it creates a most arresting display of colour. It is highly regarded in Western Australia as a cut flower.

PLATE 44: *Isopogon latifolius.*

Zoë C

LOMATIA TINCTORIA

A small genus of the family Proteaceae with eight species being found in eastern Australia and a few in Chile. The Australian species are mainly small shrubs, with varying and attractive foliage and distinctive cream flowers. *L. tinctoria* occurs only in Tasmania. *Tinctoria* (used in dyeing) refers to the powder found in the fruits which according to one source produces a rose colour.

DESCRIPTION:
Growth habit: a small, neat, erect shrub to 1 m or occasionally to 1.5 m under ideal conditions.
Leaves: the divided leaves, much like a coarse fern, are quite thick, deep green and measure up to 80 mm long. Carried quite densely on the bush, they are a feature and provide an attractive background for the pale cream flowers.
Flowers: the almost white, light cream flowers, delicately tipped with dark pink, are borne on long, erectly held sprays. They are produced over summer.
Variants: usually a uniform species displaying little variation.

CULTIVATION:
Normally an easy subject under cultivation, this species requires an average, well-drained soil, sun or semi-shade and is noted for being frost tolerant. In its natural habitat it tends to sucker, a habit not normally noted in cultivation.

PROPAGATION:
A relatively easy species to propagate, either fresh seed or cuttings of semi-ripe material being suitable.

USES:
An attractive, easily grown, smaller shrub for the garden, the foliage and flowers being appealing. Some uses for cutting, the flowers and foliage holding well in the vase.

PLATE 45: *Lomatia tinctoria.*

Zoë C

MACADAMIA TETRAPHYLLA

There are 11 species of *Macadamia*; seven in Australia and four in New Caledonia. Commonly referred to as Macadamia Nut, Queensland Nut. Famed for the delicious nuts produced which are becoming an increasingly important commercial crop in several countries. Occurs naturally in both Queensland and New South Wales, from just south of Sydney to Maryborough. Both *M. tetraphylla* and *M. integrifolia* produce edible nuts. Named after an Australian philosophist, Dr. John Macadam (1827–65).

DESCRIPTION:
Growth habit: handsome evergreen tree to 20 m in height, spreading to 7 m or more in width.
Leaves: the variable, leathery foliage, a deep glossy green, is slightly serrated and often has an undulation to the margins. Leaves can be 70 to 220 mm in length and they are borne in whorls of three or four. The new growth is a conspicuous pink or bronze colouring.
Flowers: the massed, small flowers are carried in long sprays like catkins which are from 200 to 300 mm in length. Colouring varies from pale cream or brown to an attractive pink. Flowering time is from late winter through spring. The edible nuts mature over autumn into midwinter and are encased in a particularly thick, hard, brown shell. They are carried in dense, pendulous clusters (see illustration).
Variants: selected forms are grown for features such as thinner shelled nuts and cropping performance. They are normally grafted onto strong seedling rootstocks. Seedlings are slow to crop and may not necessarily exhibit good characteristics, hence the grafting of good forms. Hybrids of *M. tetraphylla* and *M. integrifolia* have resulted in a number of recognised named cultivars of economic importance. Some of the better known *Macadamia* cultivars are 'Hinde', 'Keauhou', 'Kakea', 'Renown', 'Beaumont' and 'Own Choice'.

CULTIVATION:
Macadamias are usually grown in sheltered, warm, sunny positions in climates where frost occurrence is either minimal or non-existent. Established trees can withstand moderate frosts, but as trees produce late summer growth damage can occur from early frosts. As flowers are produced in late winter these may also be damaged by late frosts which will affect fruit set. Soils should be high in humus and well drained; the latter being particularly important to reduce the likelihood of root fungous disease such as *Phytophthora*. Pollination is important to secure good crops of nuts; for the home gardener it is usually advisable to plant two pollen-compatible varieties to ensure the cross-pollination necessary for good nut crops. Bees and moths are the main pollinating agents. The nuts are usually harvested as they begin to fall. Adequate summer moisture is an important requirement.

PROPAGATION:
Commercial cultivars are propagated by budding or grafting to perpetuate their particular qualities. Species are propagated by seed or cuttings.

USES:
Attractive trees for specimen planting, quite apart from their more obvious uses as bearers of nut crops with high commercial value. The nuts are suitable for eating raw or roasted, and are sometimes salted. They have a unique taste and the highest oil content of any commercial nut at around 77 per cent. They are also used in pastes, cakes and confectionery, and cosmetics.

Macadamia nut production first became a commercial crop in Hawaii around the turn of the century and many of the most successful hybrids between *M. integrifolia* and *M. tetraphylla* were developed there. It is now also a successful crop in its native country of Australia, parts of New Zealand, South Africa, Zimbabwe, Kenya, California and Central America.

PLATE 46: *Macadamia tetraphylla.*

Zoë C

OREOCALLIS PINNATUM
(Recently changed to ALLOXYLON PINNATUM)

Formerly known as *Embothrium wickhamii* var. *pinnata, E. pinnatum* or *Oreocallis pinnatum,* and commonly referred to as Queensland Tree Waratah. This is an appropriate name, aptly describing the appearance, but is confusing as *Telopea oreades* is also commonly known as the 'Tree Waratah'. Primarily found growing in the rainforests of Queensland and northern New South Wales in deep, well-drained soils where rainfall is plentiful.

DESCRIPTION:
Growth habit: a tall ornamental tree to 15 m or more, this distinctive species develops a sturdy trunk and robust habit, eventually forming a dense canopy of branches and foliage. In crowded situations a more erect habit develops.
Leaves: deep green, glossy, entire leaves, slightly cupped with the margins tending to curve upwards. They may measure up to 200 mm in length and 50 or 60 mm wide, and have a prominent centre rib, and become pinnately lobed with maturity.
Flowers: the rich red, tubular flowers are carried in dense, terminal clusters surrounded by foliage from spring to early summer. Borne in profusion, they create a spectacular display.
Variants: no notable variations are recorded.

CULTIVATION:
A comparatively easy species to grow in mild climates provided a good, deep, free-draining soil and a plentiful supply of moisture are available. *A. pinnatum* is tolerant of the occasional light frost, the author having successfully grown this lovely tree in such a situation. A sunny, sheltered position is preferable to produce good flowering and a well-shaped specimen. It flowers at an early stage, the author's specimen producing well at 7 years old, at which stage it had reached 3 m. In warmer climates it should perform even more quickly.

PROPAGATION:
Normally propagated from fresh seed, cuttings usually proving difficult.

USES:
A particularly handsome and spectacular specimen tree for warm climate gardens, a mature example always arousing attention and admiration. Not regarded as very suitable for cut-flower use as the blooms do not have good lasting qualities. The timber has been used commercially and is similar to that of the Silky Oak, *Grevillea robusta.*

PLATE 47: *Oreocallis pinnatum (Alloxylon pinnatum).*

PERSOONIA PINIFOLIA

Commonly known as the Pine-leaf Geebung. Most members of this quite large genus are known as Geebungs. There are some 55 species and two named hybrids. They are found in every Australian state. This species occurs throughout New South Wales and into Queensland and is particularly well known in the Sydney region, where it occurs on deep, well-drained gravel and sandstone soils. The genus was named for the South African botanist Persoon, prominent in the 18th and early 19th centuries, while the specific *pinifolia* refers to the pine-needle-like foliage.

DESCRIPTION:
Growth habit: a bushy, spreading or weeping shrub varying in height from 2 to 4 m with a spread of up to 3 m.
Leaves: the very narrow, needle-like, soft foliage is a bright green colour, and each leaf can measure as much as 50 mm in length. This striking foliage is carried densely on the branches and is a feature of the species.
Flowers: the rich yellow flowers, each no more than 10 mm long, are carried in dense, terminal clusters to 150 mm in length, mainly during summer. They droop in pendulous fashion and are always prominent. Following the flowers, large green fruit encasing the seeds are borne in spectacular terminal clusters and gradually turn a purple-brown colour.
Variants: not a particularly variable species, although some clones have been observed to produce seed which is more viable and readily germinated than others.

CULTIVATION:
A relatively easy species to grow under cultivation, most well-drained soils being suitable. Drought tolerant as well as being able to cope with wet soils in winter, this species is also fairly frost tolerant, to as low as -4 °C in the author's experience. An occasional pruning can keep the growth more compact.

PROPAGATION:
A notoriously difficult subject to propagate. Cuttings are almost an impossibility in most nurserymen's experiences, although very soft, almost wilted cuttings without top growth removed are found to work. Seed is very slow to germinate, if it does at all. One suggestion put forward is to sow seed gathered from amongst the old litter and leaf mould under mature bushes. This is reputed to germinate more satisfactorily. Also, as mentioned above, some clones naturally produce seed which germinates more readily.

USES:
A lovely specimen shrub for the garden, the weeping, bright green foliage, contrasting rich yellow flower spikes and prominent fruits all combining to make a most appealing plant. The foliage is also in demand by florists.

PLATE 48: *Persoonia pinifolia.*

Zoë C

STENOCARPUS SINUATUS

Popularly known as Queensland Firewheel Tree, Wheel of Fire, and White Silky Oak. Widely distributed along the eastern Australian coast, mainly in the rainforests of northern New South Wales and as far north as the Atherton Tableland in Queensland, but mainly nearer the coast. It is also found in Papua New Guinea. *Sinuatus* means wavy and refers to the leaf margins.

DESCRIPTION:
Growth habit: an upright, non-spreading tree which in its natural environment of the Queensland and northern New South Wales rainforests may reach 35 m in ultimate height. Under cultivation it seldom exceeds 14 m.
Leaves: the very glossy, medium-green foliage is amazingly variable. Leaves may be entire (without teeth or divisions) but still with wavy margins, others (on the same tree) may be only slightly divided, and some can be deeply and irregularly lobed and divided. They are always thick and somewhat leathery in texture and may reach 300 mm in length.
Flowers: the flowers, so much a feature of this unusual member of the protea family, are arranged like spokes in a wheel. These 'wheels' in turn may be carried in thick clusters on the branches, making a spectacular sight. Each individual flower or 'spoke' of the wheel is bright scarlet with the globular apex a deep gold. The overall effect is indeed much like a tree of firewheels. The main flowering period is from autumn into winter.
Variants: no particular variation is recorded in this uniform species.

CULTIVATION:
The Firewheel tree is a surprisingly easy species to grow, being adaptable to both light and heavy soils, sun or semi-shade and is frost tolerant to at least -4 °C once established. Soils should be not only free draining but fertile and receive adequate moisture. Generally slow-growing for the first 3 years, this species may need some pruning if a bushier habit is required and will need frost protection while young.

PROPAGATION:
Usually propagated from the winged seeds which germinate freely, but cuttings strike readily also. The latter should be taken from more mature branches, such material flowering at a younger age.

USES:
An especially beautiful, neat tree widely grown in Australia as far south as Melbourne as specimen trees, and as street trees in Sydney. Because of its neat shape it can often be grown in quite small gardens and in its juvenile stage can be used as a handsome tub plant. Although not a commercial cut-flower line, it is nevertheless often used as a spectacular flower in floral artists' arrangements, having a short but still adequate vase life.

PLATE 49: *Stenocarpus sinuatus.*

Zoë C

TELOPEA 'BURGUNDY'

As mentioned under *Telopea oreades*, natural hybrids between *T. speciosissima* and *T. oreades* occur, both in the natural environment and under cultivation. *T.* 'Burgundy' is such an example, and despite being somewhat unimaginatively named, it has appealing qualities.

DESCRIPTION:
Growth habit: a vigorous, densely bushy and erect-growing large shrub or small tree to 4 m or more.
Leaves: similar to *T. oreades* being almost entire (without serrations or undulations), deep green and prominently veined and more or less oblanceolate in shape. Typically they may measure up to 200 mm in length and 40 mm in width.
Flowers: the prominent blooms are a bright crimson shading, typical of the *T. oreades* parent. They exhibit many *T. speciosissima* flower characteristics, notably the fullness of the heads and larger size. The most intriguing aspect, however, is the variable flowering time. Frequently blooms are produced in the autumn and early winter, although late winter to early spring is the main flowering period. Stem length is good and vase life, while possibly not as prolonged as *T. speciosissima*, is better than *T. oreades*.
Variants: there are numerous variations of this hybrid theme, some not particularly notable or obvious, others showing outstanding characteristics worthy of perpetuating in cultivation. *T.* 'Burgundy' is but one of these and displays sufficient differences to make it worthy of inclusion in this book as an example of the cross.

CULTIVATION:
A particularly easy waratah to grow in the home garden, being considerably less temperamental and demanding than *T. speciosissima*. Most average, well-drained soils appear to suit this hybrid and it is tolerant of winter frosts to −10 °C in many areas.

PROPAGATION:
As with any hybrid or selected form, vegetative propagation is the method of ensuring progeny that are true to the parents' characteristics. Cuttings of semi-hardwood material are usually taken in late spring or autumn.

USES:
A worthwhile hybrid that provides a waratah for gardeners who have difficulty growing *T. speciosissima*. The ability to produce some blooms at different times of the year is another advantage, and it would appear that this habit has been inherited from a specific form of *T. oreades*. There are many similar crosses found in the wild and they often have desirable characteristics for both home garden and commercial picking purposes.

PLATE 50: *Telopea* 'Burgundy'.

TELOPEA OREADES

Commonly known as Tree Waratah, Gippsland Waratah and Victorian Waratah. First discovered and described in 1860. Naturally occurring at varying altitudes in both wet eucalypt forest and drier ridges in Victoria and parts of New South Wales, it is particularly prominent in the Snowy River area of Gippsland. *Oreades* from the Greek *oros*, meaning a mountain, refers to the habitat of this species.

DESCRIPTION:
Growth habit: appropriately referred to as Tree Waratah, *T. oreades* is considerably taller growing than its relations, reaching heights in excess of 12 m in sheltered positions with a trunk capable of reaching 450 mm in diameter. Young plants have a distinct, erect, tree-like habit, but as they mature they become branched and dense in character, assuming a large shrub-like appearance. In open positions, under cultivation, this branching habit usually becomes dominant at an earlier stage of development.
Leaves: the foliage of this species is smaller than that of the New South Wales Waratah, *T. speciosissima*. The leaves are not serrated, the margins being quite smooth, and are dark green on the upper surface and a lighter, greyish-green on the underside, where the veining is more prominent.
Flowers: flowerheads are a deeper red than those of *T. speciosissima*, having a more crimson toning, and are quite open in shape compared to the former which has tighter, often dome-shaped, flowers. The individual florets, usually numbering about 20, are held almost horizontally in a radiating pattern from the open centre. The main flowering period is during spring, and in nature specimens growing at higher altitudes may bloom a month later. The flowerheads are borne terminally on long, stout stems and new vegetative shoots will frequently emerge through the blooms before they mature.
Variants: flower colour varies in the natural habitat from deep to lighter reds, the deep crimson being more usual, and a white flowered form has been discovered quite recently, the florets having a creamy-yellow toning. A further form has exhibited the ability to flower at almost any time of year. Both in nature and under cultivation, natural hybrids between *T. speciosissima* and *T. oreades* occur and these exhibit characteristics of both species to varying degrees. A chance seedling hybrid known in New Zealand as *Telopea* 'Burgundy' frequently flowers in autumn as well as spring.

CULTIVATION:
Regarded by many gardeners as one of the easier species to cultivate, tolerating more frost than *T. speciosissima* (areas recording frosts of –15 °C succeed in growing it) and being more adaptable to varying soil types. It nevertheless requires a reasonably good, well-drained, acidic soil to grow well. Either filtered shade or full sun are suitable, and once established this species is reliable and long-lived. A yearly pruning immediately after flowering encourages more compact growth habits.

PROPAGATION:
Normally propagated from both seed and cuttings, the latter being necessary if selected forms are to be perpetuated. Grafting and tissue culture techniques may also be used to perpetuate desirable forms.

USES:
A large shrub to small tree, possibly more suited to the larger garden, it never fails to produce a spectacular display of spring bloom. Lovely for background effect or as a specimen. Also used as a stock for grafting selected *T. speciosissima* forms that may be difficult to establish or which lack vigour on their own roots. Also used for hybridising purposes. Not regarded as a commercial cut flower, *T. speciosissima* having pride of place in this respect.

PLATE 51: *Telopea oreades*.

ZoëC

TELOPEA SPECIOSISSIMA

Commonly known as the New South Wales Waratah in both Australia and other countries where it is cultivated. One of only four species all endemic to the eastern coast of Australia, they are found in New South Wales, Victoria and Tasmania. *T. speciosissima* is probably the best known Australian wildflower and is the floral emblem of New South Wales, the state where it occurs naturally over a wide area. Best specimens may be found at higher levels of the Blue Mountains growing in deep, well-drained soils with ample moisture. *Telopea* comes from the Greek *telopos*, meaning visible from afar. The specific *speciosissima* means very beautiful or splendid.

DESCRIPTION:

Growth habit: an erect, multi-stemmed, bushy shrub, varying in height from 1 to 4 m and attaining a width of 1.5 to 2 m. The bark is smooth and greyish-brown.

Leaves: oblong to wedge-shaped, usually toothed or lobed in the upper part, tapering into a long stem. Conspicuously veined, often bronze-coloured in the young, soft growth stages, but medium to dark green with maturity. A lighter shade on the undersides.

Flowers: large and prominent, carried as terminal, globular heads atop long, stout stems. The brilliant red colouring of both the flowerhead and the usually present large basal bracts creates a dramatic display during the spring months of September to November in Southern Hemisphere gardens, October being the peak flowering month in most areas. Mature bushes that have been well pruned to encourage a compact, branching habit have been recorded as bearing up to 400 blooms in a season.

Variants: considerable variation both in flowers and foliage is exhibited by seedling plants, and as a result many selected forms have been propagated and given cultivar status by nurserymen in Australia and New Zealand, these having been chosen for characteristics such as suitability for commercial cut-flower production or for brilliance of flower colour and size. Shape of flowers can vary from quite squat to compactly peaked or drawn up into an almost cylindrical shape. Some have very prominent basal bracts, while in others these may be small and insignificant. A rare white-flowered form, 'Wirrimbirra White' is in cultivation in both Australia and New Zealand.

CULTIVATION:

Not always an easy subject to establish in the home garden, *T. speciosissima* has been subjected by botanists and horticulturists to varying theories as to what constitutes an ideal garden or commercial growing environment. All agree, however, that a deep, well-drained, light soil is necessary. Some maintain that a spadeful of soil from around an old bush will introduce mycorrhizal organisms that are beneficial to successful establishment and yet other evidence suggests that this is unnecessary. There is also a conflict of opinion as to whether nutrient-rich or more impoverished soils are preferable, and as to whether some organic fertiliser is beneficial. It has been suggested that potash is important for successful cultivation, as this occurs naturally from the results of bush fires in the natural habitat. Whichever theories are correct, most gardeners seem to achieve success in average, well-drained, deep, acidic soils that are not over-fertilised. Phosphate and lime should certainly be avoided. Sunny or partially shaded situations in hot climates are preferred. Frost is not usually a deterrent to successful cultivation, waratahs being grown without difficulty in districts recording winter temperatures as low as −10 °C.

Hard pruning of telopea species immediately after flowering is essential to maintain a compact, bushy plant. Unpruned bushes soon become tall, straggly and unsightly.

PROPAGATION:

Easily propagated from the winged seeds, which should be sown fresh. Seedlings are generally true to type, but variations may occur. This species is sometimes grafted on to *T. oreades* stock. Some nurserymen propagate from cuttings of semi-hardwood taken during early summer, but success varies with different clones. In 1985 the Levin Horticultural Research Centre in New Zealand pioneered the tissue culture of *T. speciosissima* and this technique is being increasingly used for 'bulking up' improved selections.

USES:

A spectacular shrub for the home garden, providing a vivid display in spring. It is becoming a successful commercial flower crop in Australia, New Zealand, Hawaii and Zimbabwe, large areas having been planted and blooms exported to Japan and the United States in particular.

PLATE 52: *Telopea speciosissima.*

Zoë C

THE PROTEACEAE OF NEW ZEALAND

New Zealand, a small country, has only two species of Proteaceae. One was classified as coming from the genus *Persoonia,* all other species of which are endemic to Australia, but has been re-classified as a separate genus, *Toronia.* The other is from the genus *Knightia,* other species of which are endemic to New Caledonia, which is closely related to the Australian genus *Darlingia* and the New Caledonian *Eucarpha.* (These relationships illustrate the Gondwanaland theory to perfection.)

Knightia; 1 species.
Toronia; 1 species.

KNIGHTIA EXCELSA

Commonly known as Rewarewa or New Zealand Honeysuckle. One of only three species, the other two, lesser known species are from New Caledonia. Found in lowland and subalpine forests of the North Island and in the Marlborough region of the South Island of New Zealand. The generic name *Knightia* is for Thomas A. Knight, an early English botanist, while the specific Latin *excelsa* translates to lofty or high.

DESCRIPTION:
Growth habit: a tall, erect tree reaching to 30 m. Frequently described as having the habit of a Lombardy Poplar, due in part to the stiffly erect branches. Branchlets are often clothed with reddish-brown, velvety hair (tomentum). Bark is smooth and characteristically dark brown to almost black.
Leaves: stiff, light to medium green, 100 to 150 mm long, but up to 250 mm on juvenile plants and coarsely toothed. Red-brown velvet tomentum is usually seen on the underside of foliage.
Flowers: distinctive, spidery, velvety reddish-brown flower buds split open and curl back in an almost distorted fashion to reveal the stamens. These buds (50 or more) are arranged along a central stem (raceme), and the total flower mass may be up to 100 mm long and 50 mm in diameter. The main flowering period is in summer when a tree in full bloom is most impressive.

CULTIVATION:
In the garden, a close association with other trees and shrubs appears to be preferred, this resembling more closely the natural habitat and providing necessary shelter for the young establishing plant. Pruning is unnecessary as the neat shape is self-maintaining. Average, well-drained, slightly acidic soils are preferred. Heavy frost and prolonged periods of drought are both to be avoided, particularly during the establishment period.

PROPAGATION:
From the winged seed which is carried in the boat-shaped receptacles; not unlike those of the waratah *(Telopea).*

USES:
A handsome tree of neat proportions for the home garden where it has the added benefit of being attractive to nectar-seeking birds, in particular the New Zealand tui and bellbird. Both in the past and occasionally today, the attractive timber has been used for a variety of purposes. Picture frames, inlay work and parquet floors are some of the more common uses, although it has been used for furniture-making as well.

PLATE 53: *Knightia excelsa.*

Zoë C.

Zoë C

THE PROTEACEAE OF SOUTH AFRICA

South Africa is second only to Australia in the number of genera and species of Proteaceae, having 13 genera and 329 species, and the family was named after the South African protea.

This book does not attempt to describe and illustrate all the members of the Proteaceae found in South Africa, but the genera and number of species are listed below.

Aulax; 3 species.
Brabejum; 1 species.
Diastella; 7 species.
Faurea; 4 species (plus 1 in Madagascar).
Leucadendron; 79 species.
Leucospermum; 48 species.
Mimetes; 11 species.
Orothamnus; 1 species.
Paranomus; 18 species.
Protea; 82 species (plus 35 in tropical Africa).
Serruria; 44 species (plus more as yet un-named).
Sorocephalus; 11 species.
Spatalla; 20 species.

◀ PLATE 54: *Protea cynaroides.* A large autumn-flowering variant of the spectacular King protea.

AULAX CANCELLATA

Formerly known as *A. pinifolia* (alluding to the narrow, needle-like foliage). One of only three species. This genus is one of only two South African members of the Proteaceae to have male and female flowers on separate plants (i.e., flowers are unisexual). The other genus is *Leucadendron*. This species is found over a wide area of the south-western Cape at varying altitudes.

DESCRIPTION:
Growth habit: varies considerably according to locality and soil type. Under cultivation, in good soils and particularly if picked heavily, it forms a bushy shrub of 1 m. In poorer, light soils it may become a narrowly, erect shrub to 2 m. Due to the fine foliage, the overall appearance of the plant is of an open texture, although older specimens that have 'bushed up' may be thicker.
Leaves: narrow, needle-like, to 40 mm long, mid to deep green, pointed. Young growth is a soft pinkish-bronze colour.
Flowers: produced over spring months. On male plants the flowers are like erect, soft yellow, catkins. The female bears stiff, light creamy-yellow, funnel-like blooms which become dark red seed-bearing cones and persist on the bush for many months. These cones are often mistaken for buds, and are particularly attractive.
Variants: little variation is recorded apart from some minor colour differences in the cones.

CULTIVATION:
Well-drained, acidic soils are a requirement for successful cultivation, and fertilisers are generally best avoided. Frosts of −5 °C do not appear to cause damage to this species, but colder temperatures are reported to have caused damage.

PROPAGATION:
By either seed or cuttings of semi-hardwood material usually taken in autumn or late spring. Vegetative propagation is essential if a specific male or female plant is required. Seed will give a mix of both sexes.

USES:
An unusual and distinctively different small shrub suited to a variety of garden applications. The flowers, and more especially the female cones, are useful for floral art purposes.

PLATE 55: *Aulax cancellata,* female (upper left).
Aulax pallasia, female (lower right).

AULAX PALLASIA

A species inhabiting the Worcester, Tulbagh, Stellenbosch and Paarl areas in South Africa's Cape Province. Not as well known in cultivation as the more colourful *A. cancellata*.

DESCRIPTION:
Growth habit: the branches of this species grow from an underground, persistent rootstock. The shrub forms an open, fairly erect habit to about 1.5 m.
Leaves: broader, and of a lighter green shade than *A. cancellata* (see illustration).
Flowers: female flowers are a similar basic shape to the foregoing species but of a soft, lemon yellow colouring. Male blooms are carried in dense clusters and are a bright yellow. The flowering period is mid summer.

CULTIVATION:
As for *A. cancellata*.

PROPAGATION:
As for *A. cancellata*.

USES:
An interesting species to mix in a proteaceous shrub border, the foliage and flowers providing good colour contrast.

Zoë C

LEUCADENDRON ARGENTEUM

Commonly known by the names Cape Silver Tree or more usually as Silver Tree and sometimes simply Silvers. Grown by many gardeners and nurserymen in Australia, New Zealand and California from early this century. Occurs naturally on the foothills of South Africa's Cape Peninsula, in particular where soils are heavy but well drained and the trees are subject to both wind and misty, low cloud.

DESCRIPTION:
Growth habit: a small tree, usually of neat proportions, to a maximum height of 10 m. Thick, heavily foliaged branches, giving a particularly dense overall appearance. May become more sparsely branched and foliaged with maturity, particularly in hard conditions.
Leaves: up to 150 mm long and 30 mm wide. Covered with a silvery down or pubescence over the basically green colouring. The overall appearance is of a silvery-blue toning.
Flowers: on the male plant the flowers are 50 mm wide, and as they open they assume a golden-yellow colour and emit a faint scent. The colour reflects in the surrounding rosette of silver leaf bracts. On the female, the silver seed cone can vary in size, at maximum becoming a very prominent, football- or egg-shape, up to 150 mm long and 100 mm in diameter. After trees have been subjected to fire, felling or similar damage, the large, black seeds with their hairy parachutes attached are ejected from the big cone and dispersed by the wind. Male flowers are usually at their peak during spring and early summer, while the female cones may be found on trees at almost any time, although they are naturally at their flowering stage at the same time as the male. Large, fully developed cones can be seen at their best over autumn and winter.
Variants: no significant differences occur, although natural hybrids with *L. rubrum (L. plumosum)* have been observed.

CULTIVATION:
Regarded both in its native country and elsewhere as somewhat temperamental under cultivation, sudden deaths not being unusual. Although in nature this species grows on quite heavy, granite-derived soils, they are free draining, and this is an important requirement for successful cultivation. Splendid old specimens have been observed growing on light, sandy soils, volcanic and gravelly soils also appearing suitable. A life span of 10 to 20 years is the usual range under favourable conditions. Healthy specimens are usually in open positions where air movement helps to dry foliage and thus prevent fungal problems.

Generally regarded as frost tender, especially while young, *L. argenteum* can tolerate frosts of -5 °C once established but colder conditions will usually prove fatal.

Few pests trouble this species apart from the odd caterpillar attack. Heavier, damp soil conditions can cause root rot problems, however, and this is why well-drained soils are preferred.

PROPAGATION:
Although it is feasible to propagate this species from cuttings, and this is in fact the only way of guaranteeing male or female progeny, the resulting plants are usually rather spindly in appearance. Seedling plants are much bushier in habit.

USES:
A magnificent landscape feature, 'Silver Trees' are particularly beautiful when the wind is ruffling their foliage, the silver lights of the leaves showing up in splendid shimmering style. For floral purposes, the cones of the female trees are useful and dry well for permanent arrangements. Foliage is also used for background or infill in large displays. Male flowers are occasionally used, but they are not noted for their vase life.

The male flowers dry on the tree and drop off as curious little brown and silver balls, or 'hedgehogs', as our children refer to them.

PLATE 56: *Leucadendron argenteum*, female seed cones (top and left), male flower (bottom right).

LEUCADENDRON DAPHNOIDES

Occurs naturally in a small, mountainous area of South Africa's Cape at altitudes of 240 to 1000m. First botanically described in 1806. The specific name alludes to the daphne-like scent of the male flower.

DESCRIPTION:
Growth habit: quite fast growing to a maximum height of 2 m and forming a bushy shrub with a spread of up to 1.5 m. The heavily felted or pubescent younger foliage gives an overall soft, dense appearance to the shrub. Most garden specimens (with some picking or pruning) seem to maintain a height of closer to 1.5 m, and assume a rounded shape.
Leaves: up to 65 mm long and 15 mm wide. Grey-green in basic colouring, with a soft, silvery-grey, pubescent coating, particularly on new growth.
Flowers: in the male, the prominent golden-yellow central flower mass measures up to 40 mm across and is complemented by the almost pure white encircling bracts which give an overall diameter of up to 100 mm. The female flowers develop at the same time (late winter to early spring) and mature to a prominent, seed-bearing cone. In both sexes the creamy coloured bracts age to spectacular shades of orange, pink and red before reverting to green.
Variants: differences in the size of leaves and flowers have been noted and also the amount of pubescence. The example depicted in the illustration has a more prominent red margin to the bracts than is usual. This species does not appear to hybridise readily with other species in nature, the only one having been recorded being with *L. tinctum*.

CULTIVATION:
As with most leucadendrons, this species responds well to average garden conditions of good drainage, slightly acid soil and reasonably open, sunny positions. It is sensitive to fertiliser application, which should be avoided, and resents overly protected or crowded situations where fungous diseases are likely. Not a particularly frost hardy species, much more than -2 °C causing damage.

PROPAGATION:
Readily propagated from seed, the results of which may be variable, giving a mixture of male and female plants and sometimes other variations. Selected male or female forms should be propagated from cuttings of semi-hardwood material taken in late spring or autumn.

USES:
An ornamental garden plant giving a colourful display of bracts in varying shades from white to red as they mature. The neat habit, a distinctive foliage contrast with other shrubs, plus its usefulness for picking purposes all combine to make *L. daphnoides* a welcome addition to the home garden.

PLATE 57: *Leucadendron daphnoides*, male form.

LEUCADENDRON DISCOLOR

Commonly known in South Africa as Sunshine Bush or Flame Gold-tips. Occurs naturally in a restricted mountainous area of the western Cape in South Africa. Soils are usually rocky and derived from Table Mountain sandstone.

DESCRIPTION:

Growth habit: a low, bushy shrub for the first 2 or 3 years after planting, gradually forming a more upright habit and reaching a height of 2 m with a spread of 1 to 1.5 m. Readily shaped with pruning to form a lower, bushy shrub.

Leaves: mid to deep green in colour, 45 mm long and 15 to 20 mm wide. Younger foliage has reddish tints and pronounced silver tufts to the apex of the leaves.

Flowers: the typical male flower is strawberry-like in appearance and coloured bright red changing to yellow as the pollen becomes more evident. Some forms are a pale orange (as illustrated). The bracts are a soft creamy-yellow and the total width of the 'flower' when the bracts are included is up to 70 mm. Main flowering stems are not long, 400 mm being the usual, and may carry several flowers, each on its own short stem. The female has paler coloured bracts and the central seed-bearing cone is a green shade tinged with red. The main flowering period is over the spring months.

Variants: the paler coloured, male form illustrated is more tulip-shaped in the bract formation and also usually has longer stems with fewer 'multi-heads' of flowers. Other variations are recorded, size and shade being the main differences. Some interesting hybrids have been observed in the natural state and others have also been developed by nurserymen. One of the more outstanding is the result of crossing *L. discolor* with *L. gandogeri*.

CULTIVATION:

Generally an easy leucadendron to grow, preferring warm, sunny positions and well-drained, acid soils that are low in nutrients. Frost protection will be required in colder districts. Winter temperatures lower than -4°C are likely to cause damage.

Unfortunately one of the leucadendrons most subject to attack by various caterpillars. Regular spraying with a suitable insecticide from the time of bract formation will help to ensure perfect stems for picking.

PROPAGATION:

By cuttings if specific male or female or other forms are required. Use semi-hardwood taken in late spring or autumn as this is usually easy to strike. Fresh seed germinates freely but produces variations as well as a mix of male and female plants.

USES:

A showy garden plant giving good colour over spring — especially in the case of male plants, while both the male and female forms are popular with floral artists.

PLATE 58: *Leucadendron discolor*, male forms (female lower left).

Zoë C

LEUCADENDRON ELIMENSE

Originates from a small area of the south-western Cape where it occurs on mildly acidic, shallow soils overlying ironstone gravel with clay underneath. The specific name *elimense* is from Elim, the district in which the subspecies occurs. All three subspecies are regarded as rare and endangered in their natural habitat due to agricultural encroachment.

DESCRIPTION
Growth habit: a narrowly erect growth from 1 to 1.5 m with quite sparse branches typifies this leucadendron. Careful pruning from an early age can improve the branching habit.
Leaves: variable, according to the subspecies, of which there are three. The subspecies *salteri* has the largest leaves, measuring to 80 mm long and 250 mm or more in width, whereas the more typical subspecies *elimense* has foliage to 50 mm or less in length. Subspecies *vyeboomense* has leaves of less than 40 mm long and approximately 100 mm wide.
Flowers: the male plants of all three subspecies produce the more spectacular flowers and bracts, subspecies *salteri* having the largest at up to 200 mm wide overall in exceptional cases, 100mm overall being the usual. The 40-mm wide flower mass is always prominent, is produced over spring and has a characteristic strong, spicy scent, not unlike a daphne. The female flower is also strongly scented and is up to 20 mm in diameter.
Variants: as already discussed there are three distinct subspecies. These are readily distinguished by their differences in leaf and flower size, but also by different distribution patterns, although all in the same general geographic area. There is little variation within the subspecies and no natural hybrids are recorded.

CULTIVATION:
Slightly acidic soils (around pH 6) that are reasonably well drained seem to suit this leucadendron. As with most of the Proteaceae, sunny positions give the best results with bracts showing more colour than in shaded positions. It appears moderately hardy, withstanding frost of -3°C without damage.

PROPAGATION:
While seed usually germinates readily and provides progeny true to type, some variation can occur and naturally there will be a mixture of male and female plants. Cutting-grown plants provide duplicates of the parent and are essential where the specific sex or particular characteristics are to be perpetuated.

USES:
Interesting leucadendrons for mixed shrub borders. The attractive foliage in the subspecies *salteri* is a feature and this plant also provides particularly large, sunflower-like blooms (in the case of the male) which are spectacular for floral arrangements, although they do not have a good vase life.

PLATE 59: *Leucadendron elimense* (top), *L. elimense* subspecies *salteri* (bottom).

Zoë C

LEUCADENDRON GLABERRIMUM subspecies ERUBESCENS

The natural habitat of this leucadendron is on mountain slopes at altitudes of 300 to 1000 m in the western Cape in South Africa. Soils are usually sandy or stony and are derived from Table Mountain sandstone. The climate is generally hot and dry over summer and cold and wet over the winter months.

DESCRIPTION:
Growth habit: generally a bushy, compact, and fairly erect small shrub to 1 m by 1 m. The subspecies *erubescens* is usually more sprawling in habit and also has smaller foliage and flower bracts.
Leaves: basically medium green, covered with a fine, suede-like pubescence, creating an overall dull green appearance, apart from the tips and margins which are edged with red. This spectacular red colouring, so much a part of this subspecies, becomes more prominent on stems and foliage nearer the flower bracts. Leaf dimensions are 40 to 50 mm long and 7 to 9 mm wide.
Flowers: the male flower (or inflorescence as it is correctly termed) is a deep, golden yellow, fading with age. Surrounding bracts are a bright, wine-red colour, contrasting vividly with the flower. The overall diameter of flower and bract is up to 80 mm, while stem length is relatively short, varying from 200 to 400 mm. The main flowering period is from early to mid spring.
Variants: minor variations occur, mainly in the amount of pubescence on the foliage.

CULTIVATION:
Well-drained, slightly acidic soils are preferred, those of a stony or sandy nature resembling more closely those of this leucadendron's natural habitat. A moderately hardy species, tolerating frost of -5 °C once established, but not as forgiving of drought conditions as some. Under cultivation *L. glaberrimum* ssp. *erubescens* has not shown itself to be particularly susceptible to any particular pests or diseases and can be considered relatively easy to grow.

PROPAGATION:
While seed is generally easy to germinate and shows little variation, most nurserymen will probably propagate this one from semi-hardwood cuttings taken in late spring or autumn to enable quick production of strong, saleable plants.

USES:
This leucadendron is unlikely to be sought-after as a commercial cut-flower proposition, but as a showy and distinctive leucadendron for the home garden it is bound to find favour. It could have some use as a flower for smaller arrangements but the more colourful male blooms do not have a very long vase life.

PLATE 60: *Leucadendron glaberrimum* subspecies *erubescens* (male).

LEUCADENDRON GLOBOSUM

Regarded as rare in nature, this species is confined to a small area of the south-western Cape where it occurs on steep, south facing slopes at lower altitudes of 250 m, in loamy, mildly acid soils. First described by botanists in 1803. The specific name is from the Latin *globosus*, meaning spherical, a reference to the shape of the fruits.

DESCRIPTION:
Growth habit: generally narrowly erect in habit, male plants rarely exceeding 1.3 m in height with the females being taller to 2 m.
Leaves: slightly smaller and less rounded in the male plants, noticeably rounded in the female, 40 to 45 mm long on average, but sometimes to 60 mm and 25 to 30 mm wide. Light to mid-green in colour and quite prominently veined. Young foliage is covered with a soft pubescence.
Flowers: the female flower is more prominent than in most other leucadendrons, and before becoming the seed-bearing cone is an attractive golden-yellow colour up to 23 mm wide and surrounded by broad, soft yellow bracts giving a total 'flower' width of up to 70 mm. In the male, the flower is even more conspicuous, being up to 50 mm wide and a deep golden colour. Surrounding bracts are fewer and more spaced and a deeper yellow than in the female. Stems of both sexes are up to 500 mm or more in length. The main flowering period is over late spring, but flowering times will vary with localities.
Variants: no significant variants have been noted.

CULTIVATION:
Provided good drainage is available, this species is fairly adaptable to a variety of situations. Most average, acidic soils that have a reasonable level of organic matter are suitable. Light to medium frosts do not appear to cause damage. Pruning after flowering will improve the branching habit. Apart from some caterpillar damage and a tendency to attract scale insects, it is a relatively trouble-free species.

PROPAGATION:
Although seed germinates readily, most nurserymen prefer to grow plants from cuttings so that the sex is known.

USES:
A showy shrub to use in conjunction with others in massed shrub plantings. *L. globosum*, in both male and female forms, appears promising as a cut flower, although bushes do not seem to produce as many flowering stems as some other species and hybrids. It is nevertheless a particularly attractive shrub when in flower.

PLATE 61: *Leucadendron globosum*, male (left), female (right).

LEUCADENDRON LAUREOLUM

Formerly well-known as *L. decorum*. A widely distributed species throughout the south-western Cape in South Africa where it is found growing in widely divergent soil types, ranging from those derived from sandstone to others originating from granite and even limestone. Because of this, the soil pH varies from the quite acidic to mildly alkaline. Altitudes at which this species occurs are also variable, ranging from sea level to 1000 m. *Laureolum* refers to the laurel-like foliage.

DESCRIPTION:

Growth habit: usually a dense, bushy shrub to 2.5 m in height and often as wide. The thick, branching habit from ground level, with little evidence of a trunk, adds to the dense appearance. Occasionally a form will grow more erectly as in the cultivar of Hawaiian origin known as 'Colin Lennox'.

Leaves: on male plants up to 75 mm long and 15 mm wide, larger in the female, being up to 95 mm long and 17 mm wide. Medium green in colour, noticeably veined and covered on both sides with a fine, short, velvety pubescence. These minute hairs tend to give the foliage a hint of silver in some lights. It is more noticeable on younger growth on the female plants.

Flowers: it is the colourful, golden yellow, terminal leaves or bracts, referred to as 'flowers' by most gardeners that give the bushes the bright appearance over the mid winter months. The female bracts are among the largest of leucadendrons, the total width of the terminal 'flowerheads' being up to 100 mm and as long. The female bracts gradually open to display the prominent seed-bearing cone which may remain unopened on the bush for several years until the demise of the shrub through fire or other reasons causes the cones to open and the seeds to be expelled. Colouring of the bracts ranges from pale lemon to rich gold with intermediate shades according to the cultivar. Plants grown in higher rainfall areas with warm, humid conditions will often not colour as well as those in cooler, drier climates. Male plants usually have very bright gold bracts which surround the pollen-bearing flower. They are normally somewhat smaller than the female.

Variants: it is interesting that in its natural habitat *L. laureolum* is regarded as a constant species, not showing any degree of variation and not hybridising to any extent. For some reason, when cultivated, this species gives considerable variety; in growth habit, stem length, bract size and colouring and even, to some extent, flowering time. This has given rise to a number of different cultivars being offered by nurserymen, especially in New Zealand, Australia and more recently in Hawaii. Names such as 'Rewa Gold', 'Colin Lennox', 'Waverley Gold' and 'Royal Gold' are just some that have been coined to differentiate them.

CULTIVATION:

Usually regarded as an easy leucadendron to grow in most well-drained sunny positions. As outlined above, this species, due to the varying soil types in the natural habitat is very adaptable. However, excessively rich soils can cause lush growth which is more susceptible to frost damage and disease and in such conditions bract colour is often pale. It is reasonably frost tolerant, the author having seen examples growing in areas receiving winter frosts of −15 °C. Regarded as a long-lived species and retains a good shape providing some pruning is carried out after flowering. Apart from occasional caterpillar damage *L. laureolum* is not subject to any particular pest or disease problem.

PROPAGATION:

While seed germinates readily, to obtain the desired male or female plant or to ensure that the plant will be true to type, propagation should be undertaken from cuttings only. These are usually taken from semi-hardwood in autumn, but some success can be achieved from material taken in late spring as well.

USES:

A deservedly popular leucadendron for both colourful winter display in the home garden and for commercial picking purposes.

PLATE 62: *Leucadendron laureolum*, male (lower left), female (top and right).

Zoë C

LEUCADENDRON 'MAUI SUNSET'

A significant hybrid of *L. laureolum* x *L. salignum* parentage originated by an Hawaiian nurseryman. A similar type of cross to *L.* 'Safari Sunset', but using a yellow form of *L. salignum* instead of the red.

DESCRIPTION:

Growth habit: a fairly erect, compact habit typifies this hybrid. It does not have the vigour of hybrids of similar parentage such as *L.* 'Safari Sunset, *L.* 'Red Gem', and *L.* 'Silvan Red', being slower growing and more open in form.

Leaves: deep green, with a red blush towards the apex, margined with red in younger growth. New leaves carry a distinct pubescence along the margins, culminating in a silver tuft at the leaf apex. Older foliage becomes a deeper grey-green with smooth margins but retains a small tuft of silver hairs at the apex. The leaves are 65 mm long and 10 mm wide.

Flowers: the involucral bracts which enclose the female flower are particularly striking, the soft greenish-white colouring giving the overall impression of a white *L.* 'Safari Sunset'. The cream bract colouring persists through the spring months, gradually becoming suffused with soft pink shadings as summer approaches. The overall width of the bract formation is up to 60 or 70 mm and slightly longer. Stem length is medium at 300 to 600 mm.

CULTIVATION:

The basic conditions required by *L.* 'Safari Sunset' and similar leucadendrons are suitable for *L.* 'Maui Sunset'. These are free-draining, acid to neutral soils, open, sunny positions and winter temperatures not exceeding −6 °C. In cold areas where plants have 'hardened' up before hard frosts occur, hybrid leucadendrons of this type have withstood occasional frosts of −12 °C. Pruning in late spring will encourage a bushy growth habit and longer flowering stems.

PROPAGATION:

As with all cultivars, propagation should be from cuttings.

USES:

As a garden plant *L.* 'Maui Sunset' has proven popular for its neat, compact habit and for producing its distinctively coloured bracts at a time when many other leucadendrons have finished. For this same reason it has found favour with some commercial growers seeking to extend their season. Floral artists enjoy using *L.* 'Maui Sunset' for the subtle colourings which blend well with other flower material.

PLATE 63: *Leucadendron* 'Maui Sunset'.

ZoëC

LEUCADENDRON 'PISA'

A particularly unusual and significant cultivar, originating at the nursery of G. Jewell, Rewa, New Zealand, from imported seed of *L. floridum*. Now being cultivated in several countries.

DESCRIPTION:
Growth habit: a vigorous cultivar, quickly forming a densely erect shrub of up to 2.5 m in height with a spread of 1 to 1.5 m.
Leaves: grey-green, this colour being imparted by the covering of fine, silver hair. The distinctive leaves are 65 to 70 mm in length, 6 to 7 mm wide, and have a very characteristic twist.
Flowers: prominent, lime-sulphur bracts surround the large, silver-coloured, seed-bearing cones which are considerably larger than in the usual *L. floridum* species. The overall width of the bract formation is up to 90 mm. Each bract formation is carried on a short stem of only 80 to 100 mm, but many are found on each flowering branch, these being a metre or more in length. Flowering time is usually mid to late spring.

CULTIVATION:
Easily grown in most soils provided they are acid, free draining and not too high in nutrient levels. Open, sunny positions assist good bract colouring and this hybrid is particularly wind-tolerant in open, exposed positions. It is also surprisingly frost resistant, instances of regular frosts of –15 °C having been tolerated in situations where the plants were 'grown hard' in an old riverbed site. Very few pests or diseases appear to affect this leucadendron, apart from some caterpillar damage and minor attacks of white scale.

PROPAGATION:
As with most cultivars, propagation should be from cuttings of semi-hardwood taken in late spring or autumn, the latter usually proving best.

USES:
L. 'Pisa' provides excellent foliage contrast in the home garden, as well as providing delightful, long-lasting material for flower arrangements in the spring. Floral artists and commercial flower growers have found it useful also.

PLATE 64: *Leucadendron* 'Pisa'.

Zoë C

LEUCADENDRON PROCERUM

A widely distributed species in inland areas of the north-western Cape in South Africa, where winters may be cold and summers hot and dry. Main rainfall occurs over winter. Soils are generally well drained, acidic and low in nutrients. First discovered and described by botanists in the early 1800s.

DESCRIPTION:
Growth habit: this tends to be a variable species; in its natural habitat it is an erect-growing shrub to 2 or even 3 m. Under cultivation the tallest specimen the author has seen was in the region of 1.8 m, while some forms appear to be lower growing and bushy to 1 m by 1 m.
Leaves: this species has distinctive blue-green foliage which, particularly when immature, has silvery white hairs along the margins and tufted at the leaf apex. Usual measurements are 50 mm long and 10 mm wide. Some cultivars have smaller leaves.
Flowers: the male flower of this species is rather like a smaller version of *L. discolor*, having the orange, strawberry-like inflorescence, but is surrounded with ivory-white bracts. The female flowers are a typical hard, green cone with broader white bracts. Overall width of the bract formation of both plants is 50 to 60 mm. Flowering time is mid spring.
Variants: as mentioned under growth habit, in cultivation this is a variable species. One cultivar, sold in some countries as *L.* 'Cherry Glow', is possibly a hybrid, being considerably smaller than the type and having more colourful red male flowers and yellow bracts. It rarely exceeds 1 m in height and spread.

CULTIVATION:
Relatively easy to grow in average, well-drained soils in sunny positions. Caterpillar damage can be a slight problem and some forms appear less frost hardy than others. The author has found the female form illustrated to be particularly frost sensitive, much more than −2°C causing damage to developing bracts. The cultivar *L.* 'Cherry Glow' appears frost tolerant, however.

PROPAGATION:
Particularly in view of the variations discussed above, propagation should be from cuttings to ensure particular forms are perpetuated. On the other hand, seed usually germinates readily and some interesting variations can occur.

USES:
A delightful garden plant for its neat habit and attractive flower bracts. All forms are useful in floral art, particularly for posy bowl work.

PLATE 65: *Leucadendron procerum*, male (top and right), *Leucadendron procerum*, female (top left), *Leucadendron* 'Cherry Glow' (lower left).

Zoë C

LEUCADENDRON RUBRUM

Formerly *L. plumosum*, this unusual species has a particularly wide distribution throughout the mountains of the south-western Cape in South Africa, where it is found growing at altitudes of 250–1300 metres above sea level on light, free-draining soils.

DESCRIPTION:
Growth habit: a neat, erect-growing bush of 1.5 to 2 m or slightly more, the male plants more densely branching than the females.
Leaves: grey-green, quite narrowly lanceolate, and usually 70 mm long and 8 mm wide, although some forms are recorded as having longer leaves.
Flowers: produced over late winter and early spring. Male plants produce large numbers of small, fluffy, light yellow flowers (not depicted in our illustration), while females produce unusual, cone-like, greenish yellow flowers with the stigmas grouped in a protruding tuft from the tip. This cone gradually enlarges and assumes an attractively multi-coloured orange, green and brown colouring. This contains the seeds which are not usually released for several years, and only then when the bush dies from fire or other causes. Female flowers may measure up to 40 mm in length, this increasing to 50 mm as they become seed cones. An open cone, at the seed dispersing stage may measure 60 mm across. These three stages are illustrated.
Variants: in nature some differences in female cone colouring can be seen, and plants grown from seed often show this variation. In South Africa, natural hybrids between *L. rubrum* and *L. argenteum* have been recorded.

CULTIVATION:
Regarded as a comparatively undemanding species to cultivate, *L. rubrum* is best suited to growing in areas where soils are light, well drained, and moderately acid. It will produce better cone colouring in open, sunny positions. In most areas it seems to tolerate midwinter frosts of up to -8 °C. It is not particularly prone to caterpillars or any other pests or diseases.

PROPAGATION:
Seed germinates readily, but if particular forms, such as male or female or any variants, are required, cuttings of semi-hardwood material should be taken in autumn.

USES:
An unusual leucadendron for the collector or floral artist seeking unusual cones suitable for use either fresh or dried. Not a showy garden plant, although the grey foliage colour provides appealing contrast among other green-leaved shrubs.

PLATE 66: *Leucadendron rubrum*, female (opened dry fruit, lower right).

LEUCADENDRON 'SAFARI SUNSET'

Outstanding hybrid originating from hybridisation between selected forms of *L. laureolum* and *L. salignum* (red form) by Ian Bell and the late Jean Stevens of Wanganui, New Zealand in 1961–2. Regarded internationally as the standard to achieve in hybrid leucadendrons, and undoubtedly the most widely planted leucadendron for commercial cut-flower purposes.

DESCRIPTION:
Growth habit: an exceptionally vigorous hybrid, quickly reaching 2.5 m or more in height if left unpruned. Densely erect in habit, more so with pruning. Stem length can exceed 1 m.
Leaves: to 90 mm in length and 15 mm wide. Deep green with a distinctive wine flush particularly evident nearer to the involucral bracts.
Flowers: insignificant in themselves, the female cones of this hybrid are surrounded by the famed, large, colourful bracts of an arresting deep wine-red shade when at their best over early winter. At that time of year, the bract formation may measure 100 mm across or 150 mm plus when 'reflexed', as some growers do by flattening the bracts down while still soft on the bush to give an open, spreading effect to the 'flower'. The red bract colour usually commences during late summer, reaching a peak with the firming of the bracts in autumn. As winter progresses, the bracts open to display a light golden-yellow shade which spreads outwards from the flower cone. At this stage the flower bracts are sometimes marketed by commercial growers under other names such as *L.* 'Tricolor' and *L.* 'Rainbow'. At this colour stage they have exceptional vase life — as much as 2 or 3 months. The length of the flowering stems is also exceptional, 1 m being quite commonplace.
Variants: as there were approximately 18 seedlings from the original cross and all flowers were marketed at some time, it is regarded as likely that there may be more than one form of *L.* 'Safari Sunset' being sold today. This is due to nurserymen taking propagating material from cut-flower markets and not realising that the stock plants exhibited different characteristics from one another, being seedlings. A similar cross carried out by Proteaflora Nurseries in Australia is known as *L.* 'Silvan Red'.

A recent development are variegated forms developed by two nurseries and named *L.* 'Jester' and *L.* 'Safari Sunshine', the former is a registered cultivar.

CULTIVATION:
Most well-drained soils are suitable. Although this hybrid is remarkably tolerant of varying soil types, from acid to alkaline and poor to rich, ideally a soil that is mildly acidic and not excessively high in nutrients is preferred. In exposed, windy positions, larger bushes will need staking or some form of support to prevent them being blown over. Sunny positions will give best bract colour. In cooler areas, *L.* 'Safari Sunset' is usually quite hardy, tolerating midwinter temperatures as low as -12°C, but in milder areas, heavy early frosts have caused serious damage.

Until recently caterpillar damage was the only pest or disease of significance to worry commercial growers, but the 'Silver Leaf' disease *(Chondrostereum purpureum)* has now become a major threat. Diseased plants should be burnt and old stumps of willow, poplar, peaches and apples which harbour the fungus should also be destroyed to prevent further infestation.

PROPAGATION:
By cuttings only, taken from semi-hardwood, usually in autumn, but may be taken at other times if the wood is suitable.

USES:
Multitude. The long flowering season, vase life, vigour and adaptability have all combined to make this leucadendron unquestionably the most outstanding to date. As a commercial cut-flower proposition it has more going for it than any other, large quantities being exported from New Zealand, and more recently Australia, to Northern Hemisphere markets. As a colourful and versatile garden shrub it is certainly outstanding for all but the coldest areas and it has even been used successfully as a hedging plant.

PLATE 67: *Leucadendron* 'Safari Sunset'.

LEUCADENDRON SALIGNUM

This species is the most widely distributed leucadendron, being found throughout the south-western Cape in South Africa at altitudes ranging from near sea level to mountainous terrain, in climates from cold to warm and humid and in soils of widely differing types. Commonly known in its native country as Geelbos. Originally incorrectly known as *L. adscendens*.

DESCRIPTION:
Growth habit: not particularly variable in growth habit in its natural habitat, except where fire has caused regeneration from the persistent rootstock when a lower coppiced effect results. Normal growth both in the wild and under cultivation is in the region of 1 to 2 m depending on both conditions and the particular form. Growth is generally described as bushy and erect. Further growth variations are described below under 'Variants'.
Leaves: in the typical *L. salignum*, foliage measures up to 60 mm long and 10 mm wide, is usually a deep green colour and frequently tipped dark red.
Flowers: both female and male forms have colourful bracts, the males having the bright yellow, pollen-bearing, central flower mass. Bracts are more usually varying shades of yellow in most forms seen in the natural habitat, but red bracts are not uncommon. Colouring commences in late summer with bracts assuming deepest colours over winter. Maximum width of the flower formation, bracts included, can reach 100 mm.
Variants: as already mentioned, the effects of veld fires can cause some variation in growth habit, and both red and yellow forms as well as variants with wider leaves can be found in the natural habitat of this species. In cultivation,however, a wider range of forms can be found under a bewildering array of names. These range from a prostrate form with cream bracts, known as 'Incisum', through a deep red semi-prostrate one known as 'Red Carpet', to more typically bushy examples with red bracts known by names such as 'Mrs Stanley', 'Manakau Red', 'Red Devil' and 'Fire Glow'. In the bushy forms with varying shades of yellow bracts there are examples such as 'Early Yellow', 'Yellow Devil', 'Durban Gold', 'Late Yellow' and 'Yellow Bird'. A recent hybrid introduction, 'Inca Gold' (*L. salignum* x *L. laureolum*), exhibits more *L. salignum* characteristics but has larger flower bracts of yellow shaded with red.

CULTIVATION:
One of the easiest leucadendrons under cultivation, *L. salignum* is adaptable to varying soil types so long as drainage is reasonable. Most variants have proven frost hardy, are able to withstand wind in more exposed situations and are long-lived, reliable garden plants. They are not normally susceptible to pests and diseases, the occasional caterpillar attack being the only problem normally encountered.

PROPAGATION:
Although an easy species to grow from seed, most nurserymen will propagate from semi-hardwood cuttings taken in late spring or autumn. This ensures the desired characteristics of selected forms are perpetuated.

USES:
All cultivated forms are useful and reliable garden plants, giving colour over an extended period during autumn and winter. Most are also popular as cut flowers, being particularly useful for smaller arrangements and posy bowls.

PLATE 68: *Leucadendron salignum*, typical red form.

Zoë C

LEUCADENDRON SESSILE

Formerly known as *L. venosum*, this species is found growing naturally over a wide area of the south-western Cape in South Africa at altitudes ranging from near sea level to 500 m in more inland areas. Soils are generally of heavy, clay types derived from granite, but other types are also evident in the habitat. Wherever this species is found, the rainfall is high and mist and cloud commonplace.

DESCRIPTION:
Growth habit: a dense, bushy shrub to no more than 1.5 m as a rule. Sturdy branches are usually heavily clothed with the neat foliage. A short trunk is evident before the branches begin, and the overall appearance is nearly always neat and compact.
Leaves: male plants have slightly smaller leaves than the females, measuring to 60 mm long and 15 mm wide. Colour is a relatively light, bright green with a reddening towards the tips. Young growth is flushed with bronze tonings.
Flowers: male flowers are prominently displayed as broadly rounded, golden-yellow buttons which open to form a fluffy mass with large amounts of pollen. They are attractively set off by the surrounding clear yellow bracts which are such a feature of the bush. The female bracts are longer, but a softer shade of yellow, and surround the prominent cone. The main flowering time is early spring.
Variants: some variation occurs, bract colour and shape being the main areas to show differences. Some variants have been given cultivar status.

CULTIVATION:
An adaptable species as soil type is not critical. Light, sandy soils to heavy clays grow *L. sessile* equally well. Reasonably frost hardy, mid winter temperatures as low as -6°C not causing damage. Adequate moisture throughout the year is a requirement. The only pests or diseases noted to any extent with this species are caterpillars and white scale.

PROPAGATION:
As with most leucadendrons seed germinates readily, but provides a mixture of male and female plants as well as other variations. For this reason nurserymen propagate from cuttings.

USES:
A lovely, neat, brightly coloured shrub for late winter and early spring colour in the garden. Some use in floral art, but not regarded as a suitable species for commercial flower growing.

PLATE 69: *Leucadendron sessile*, male form.

LEUCADENDRON TINCTUM

A quite widely distributed species throughout the south-western Cape, where so many leucadendrons are found in South Africa. Mainly grows on lower mountain slopes on acid soils derived from Table Mountain sandstone. Formerly widely known incorrectly as *L. grandiflorum*. Commonly known in South Africa as Rose Cockade.

DESCRIPTION:
Growth habit: a low, bushy and compact shrub of 1 to 1.5 m in height with an equal spread. Sturdy branches often curve downwards near the base of the bush before sending up vertical growth.
Leaves: quite large leaves are up to 110 mm long and 25 mm wide, although smaller examples of 70 mm long are probably more usual, especially on older specimens. Colouring is variable, from deep green with a red margin to blue-green shadings.
Flowers: the female flower is small, yellow, cone-like and inconspicuous until it matures into the large, red, seed-bearing cone of up to 30 or 40 mm in diameter. This colour is repeated in a spreading flush into the surrounding bracts, and it is this colouring that gives the bush its common name Rose Cockade. The red colours of both cones and bracts persists for several months over late spring and summer. The actual flowering is over winter and it is then that the bract colouring is usually a soft yellow shading.
Variants: bract colouring is particularly variable, ranging from soft yellows and limes through deep golds to red.

CULTIVATION:
If well-drained, acid soils are provided, this leucadendron is usually adaptable to most garden situations. Better cone and bract colour will be achieved in open sunny positions. Few pests and diseases affect this species in most regions.

PROPAGATION:
As seedling plants show considerable variation, it is usual to propagate this leucadendron from cuttings. These are taken from semi-hardwood in autumn.

USES:
A particularly attractive shrub for the home garden, its compact habit and extended season of flower and cone colour assuring it of a place in proteaceous sections of the garden. Insufficient stem length for commercial purposes but delightful for the floral artist's creativeness.

PLATE 70: *Leucadendron tinctum,* female form.

Zoë C

CONE-BEARING LEUCADENDRONS

Female leucadendrons produce seed-bearing cones, and in many cases these are more spectacular than either the flowers or bracts. Three distinct examples are illustrated and described.

LEUCADENDRON XANTHOCONUS HYBRID

A hybrid of *L. xanthoconus* and *L. salicifolium* with uncertain origins.

DESCRIPTION:
Growth habit: a large, bushy shrub to 3 m.
Leaves: narrow, 60 mm long, 6 mm wide, mid-green, sharply pointed.
Flowers: male flowers are small and surrounded by soft yellow bracts. The female flowers develop into quite large seed-bearing cones which turn a bright red shade as summer approaches. Borne on long, sturdy branchlets.

CULTIVATION:
Generally an easy species to grow, most well drained, acid to neutral soils being suitable. Tolerant of moderate frosts, to at least −6 °C. A sunny position is necessary to ensure maximum cone colouring.

USES:
Too large for most home gardens, but popular with florists, the cones lasting well, and the good stem length being particularly useful.

LEUCADENDRON GALPINII

Found growing in sandy soils near the coast in the southern Cape of South Africa.

DESCRIPTION:
Growth habit: an erect bush to 2 m or less.
Leaves: narrow, slightly twisted, up to 50 mm long by 4 mm wide. Silvery, grey-green in colour. A distinctive feature of the bush.
Flowers: inconspicuous, the most noticeable being the attractive silver cones produced by the female. These persist on the bush for some time and may measure 22 mm in diameter. Seen over late spring and summer.

CULTIVATION:
Easily grown in most well-drained soils and not subject to pests.

USES:
Attractive for home garden display, the silver foliage providing good contrast. Cones popular for floral artistry.

LEUCADENDRON MACOWANII

Considered rare in nature. Found in a limited area of the Cape Peninsula in moist soils near streams.

DESCRIPTION:
Growth habit: large, bushy shrub to 2 m with an equal spread.
Leaves: larger in the female, up to 80 mm long, 10 mm wide. Deep green.
Flowers: deep gold in the male with green bracts, inconspicuous female flowers mature into prominent, large seed cones to 50 mm long and 30 mm in diameter, shaded silver and reddish brown, at their peak over summer, while flowering occurs over midwinter.

PLATE 71: *Leucadendron xanthoconus* hybrid, female cones (centre), *Leucadendron macowanii*, female cones (bottom left), *Leucadendron galpinii*, female cones (right).

Zoë C

CULTIVATION:
Contrary to expectations in view of the natural habitat, this species thrives in average, well-drained soils and does not appear particularly demanding under cultivation.

PROPAGATION:
All three of the foregoing species will need to be propagated from cuttings if the cone-producing female plants are required. Seeds, which usually germinate readily, will give a mixture of male and female plants.

USES:
Not likely to be a popular garden plant, but the spectacular cones are useful for floral purposes. They also dry well.

LEUCOSPERMUM CATHERINAE

A distinctive species found growing at higher elevations of 650–1300 m in inland areas of the western Cape of South Africa. Soils are generally sandy or rocky, and moist situations near streams are favoured. Winters may be cold. Commonly referred to either as Catherine's Pincushion or Catherine Wheel Leucospermum.

DESCRIPTION:
Growth habit: a dense, bushy shrub of up to 2.5 m in height with an equal spread. Under ideal conditions it may attain the larger proportions of 3 m. It generally produces a short, thick main trunk at the base of the bush.
Leaves: up to 100 or 120 mm long, and 20 to 25 mm wide, grey-green in colouring and either three or four-toothed at the leaf tips. The tips of the 'teeth' are red, and this colour may show also as a light margin along leaf edges near the ends.
Flowers: a pale orange colour, deepening in intensity with age. The tips of the styles are a bright mauve-pink shade, contrasting against the uniform colour of the rest of the bloom. Main flowering time is from late spring to mid summer. Overall width of the open flowers may be up to 150 mm diameter, while stem length is good at 400 mm or more.
Variants: no particular variants are described in South African literature, but plants grown from seed have been noted to vary in colour intensity.

CULTIVATION:
Regarded as an easy leucospermum to cultivate, *L. catherinae* appears to adapt to almost any well-drained, acid soils in sunny, open positions. It may be seen growing happily in sandy beach sections or inland in well-drained river silts. Once established, this species seems to be able to tolerate some frost, although inland regions experiencing heavy frost may encounter problems. Apart from occasional caterpillar damage to foliage, this leucospermum is little troubled by pests or diseases.

PROPAGATION:
Generally, plants raised from seed appear to be true to type, apart from slight variation of colour intensity. Most plants sold today are grown from cuttings taken in early autumn, using semi-hardwood material.

USES:
As a larger bush with a striking flower display over late spring and early summer, *L. catherinae* is very worthwhile, being easy to grow, reliable and long-lived. For picking purposes, this species may have limited potential only. The flowers are probably too large and bulky for export packing, and the colour too pale. For local market use and in dramatic floral arrangements, however, *L. catherinae* has proven to be popular and has a good vase life.

PLATE 72: *Leucospermum catherinae.*

Zoë C

LEUCOSPERMUM CONOCARPODENDRON

A widely distributed leucospermum throughout the south-western Cape and one of the first to be described by early botanists. It grows on a wide range of soil types and in varying climatic regions. Soils can be heavy clays or sandstone types with variations between these two extremes. However, they are always well-drained, acid, and are usually on north or west facing slopes.

DESCRIPTION:
Growth habit: in nature may form a large, rounded shrub or small tree of up to 4 or 5 m in height, but under cultivation it is rare to see shrubs reaching more than 2 m, and they maintain a neat, compact habit with dense branches arising from a short trunk.
Leaves: up to 110 mm long and 50 mm wide, with varying numbers of prominent teeth at the apex. The grey, densely hairy character of the foliage distinguishes this leucospermum from its subspecies *viridum* which has deep green, glabrous leaves.
Flowers: compact, dense flowerheads of an arresting deep yellow are borne over early to mid summer. The heads may measure 100 mm in diameter when fully developed and are long-stemmed at up to 400 mm.
Variants: As mentioned under 'Leaves' (above), there is a subspecies, *viridum*, with deep green, glabrous foliage. It is widespread. A hybrid between *L. conocarpodendron* and *L. cuneiforme*, raised in Hawaii, is known as *L.* 'Hawaii Gold'.

CULTIVATION:
Generally easy to grow in most well-drained, acid soils. An open, sunny position is preferred, and sloping sites that allow frost to drain are desirable as this species may be prone to damage in heavy frosts. As with most of the leucospermums, the occasional caterpillar can cause foliage damage, but this is controlled by using a good insecticide. No particular diseases will give any cause for concern as a rule.

PROPAGATION:
Propagation from seed may give variable results, natural hybrids sometimes appearing in the progeny. The usual method used is cuttings which are taken from semi-hardwood in late summer or early autumn.

USES:
A showy shrub for early summer colour in the home garden, the neat compact habit enabling it to be accommodated in shrub borders with ease. One of the best yellow-flowered leucospermums for picking purposes, the dense, deep yellow flowers on good stems making for a desirable cut flower for both local market and export. Also popular with floral artists.

LEUCOSPERMUM CUNEIFORME

Formerly widely known as *L. attenuatum*, this species is superficially similar to *L. conocarpodendron*, flowering at the same time. Flowers are a lighter yellow and more open in form. Leaves are longer and more softly pubescent, especially while young.

PLATE 73: *Leucospermum conocarpodendron* (top), *Leucospermum cuneiforme* (bottom).

LEUCOSPERMUM CORDIFOLIUM

L. cordifolium, in all its diverse colour and growth forms is to be found growing over a broad distribution area of the south-western Cape in South Africa. It is found only on acid soils that are derived from weathered Table Mountain sandstone. Rainfall occurs over winter and ranges from 600 to 1150 mm per annum. The terrain is generally open and hilly and bushes occur either individually or in dense stands of several hundred specimens. Formerly known as *L. nutans*. As with most leucospermums, referred to as 'pins' in the cut-flower trade.

DESCRIPTION:
Growth habit: *L. cordifolium* is a very variable species, and may be found occasionally as an almost prostrate plant or a taller, vigorous bush of up to 3 m in height with an almost equal spread. Typical specimens of the forms cultivated may reach 2 m in height with an equal, bushy spread.
Leaves: variable from one cultivar to another. Difficult to be precise in a description because even on one branch of a plant leaf shape may change progressively from distinctly toothed or notched leaves on the lower parts to entire, rounded leaves towards the tip. Where leaves are notched at their apex, they are usually three-pointed, each tooth being tipped with dark red. Size is variable, from 20 to 80 mm long and 20 to 45 mm wide. Colour is usually a deep green, often with greyish overtones. The overall appearance of the foliage is of a dense, stem-clasping mass of greenery.
Flowers: most *L. cordifolium* variants commence flowering in late spring and carry on through the midsummer period. Some, especially some of the newer hybrids, may start up to a month earlier than the usual types. Colour may range from soft yellows through pale orange shades to the popular, intensely deep orange-red colours. Tendency to multi-heads is common in *L. cordifolium*, more so in some forms than others. The cultivar *L.* 'Harry Chittick' is a good example of this habit, rarely throwing single flowerheads, and up to four and even five heads per stem not uncommon. Stem length is variable also, ranging from as short as 300 mm to as long as 1 m. The size of individual flowerheads is also variable, but 100 mm is an average diameter, with some 'monsters' reaching 200 mm.
Variants: There are now so many different named forms and hybrids of this species that it would fill a page or more to adequately list and describe them. Suffice it to say that most are worthy of cultivating and improvements are continually being introduced. New Zealand probably led the way in the 1960s with named selections, but the species' homeland, South Africa, now leads the way, especially with hybrids, followed closely by Hawaii.

CULTIVATION:
Regarded as one of the easier leucospermums to grow. Most well-drained, acid soils are suitable, and it may be seen thriving in sandy beach conditions or in heavier silt soils in more inland situations. Full sun is an essential requirement to ensure plentiful blooms of good colour. Frost tender while young, it is able to withstand medium frost once established, although some cultivars are more frost hardy than others. Caterpillar damage may occur occasionally and a reliable insecticide will need to be used. Tip dieback may also occur infrequently, and this disease can be controlled with a systemic fungicide.

PROPAGATION:
As it is a variable species when grown from seed, it will be necessary to take cuttings in order to perpetuate the characteristics desired. These may be taken as wood approaches the semi-hard stage in late summer.

USES:
A delightfully spectacular shrub for summer colour in the garden, and very useful for dramatic floral arrangements. Flowers have a good vase life, but commercial growers experience difficulties with flowerheads breaking off stems while being shipped.

PLATE 74: *Leucospermum cordifolium*, cultivars.

LEUCOSPERMUM 'FIREFLY'

A particularly notable and distinctive hybrid resulting from crossing *L. tottum* with *L. vestitum*. Originating in South Africa, this hybrid is now widely grown in other parts of the world, including Hawaii, Australia and New Zealand, where it has become a popular garden plant. Both parent plants are found growing in similar regions of the south-western Cape in inland mountainous areas on sandstone soil types. Because of cold winters and summer droughts both parents, and therefore this hybrid also, are regarded as hardier than coastal leucospermum species.

DESCRIPTION:
Growth habit: a particularly compact, neat, bushy growth typifies this hybrid leucospermum. After five years' growth it will be unlikely to exceed 1 to 1.5 m in height with a similar spread.
Leaves: distinctively grey-green in colouring, 50 to 55 mm long and 15 mm wide. Mature leaves have three prominant notches at the apex, each tipped with red. Flowering stems may exhibit smaller foliage without the notches as the foliage nears the flowerheads.
Flowers: compact, neat and very rounded. Up to 100 mm across when fully developed, and 50 mm deep. Colouring is an orange-red which takes on a deeper, glowing shade with age. Stem length is quite short at 200 to 300 mm. The main flowering period is over early summer.

CULTIVATION:
An adaptable shrub, growing happily in both deep silts and light sandy soils as well as heavier clay types. Tolerant of midwinter frosts to at least -5 °C and rarely affected by drought once well established. Sunny positions give richer flower colour. It does not seem to be particularly affected by pests or diseases in most localities.

PROPAGATION:
As with all hybrid plants, *L.* 'Firefly' will need to be propagated vegetatively (from cuttings) to ensure the progeny is true to type. Cuttings should be taken over late summer or early autumn when wood is approaching the semi-hardwood stage.

USES:
This hybrid leucospermum is proving popular with home gardeners for its compact, neat growth habit and showy, intensely red flower colouring. It may have some potential as a cut flower on the local market, but is too small and short-stemmed for export purposes.

PLATE 75: *Leucospermum* 'Firefly'.

LEUCOSPERMUM GLABRUM

Found growing in a restricted area of South Africa's southern Cape where it occurs almost exclusively on cool, south-facing slopes in moist, peaty soils. Rainfall averages 1000 mm per annum and is evenly distributed.

DESCRIPTION:
Growth habit: an erect shrub of up to 3 m in height, sometimes attaining the proportions of a small tree with age.
Leaves: bright green, glabrous leaves, up to 120 mm long and 50 mm wide. Prominently toothed at the tips, having anything from seven to 14 teeth, each distinctly tipped with red.
Flowers: large, measuring up to 100 mm in diameter, and characterised by the brilliant perianths or 'ribbons', as they are popularly and appropriately known, of rich carmine red. Although the usual form has orange styles, a form with yellow styles is not uncommon (both are illustrated). Stem length is 500 mm or more and the main flowering season is over early spring, and occasionally summer in some areas.
Variants: as mentioned above, both the colour of the styles and the flowering time is variable. Some interesting hybrids have evolved from this species, one of the earlier ones being *L.* 'Veldfire' which is similar to *L. glabrum* but more yellow in colour and has a deep green, more heavily toothed leaf. *L.* 'Scarlet Ribbon' is a newer, striking hybrid resulting from a cross between *L. glabrum* and *L. tottum*. It is mostly salmon pink in colouring with the characteristic red 'ribbons' creating a vivid contrast.

CULTIVATION:
Contrary to the natural habitat where soil conditions are cool, moist and peaty, *L. glabrum* thrives under cultivation in very well-drained, light and sandy soils in sunny, warm positions. It has shown itself to be a very adaptable leucospermum and of easy cultivation. Few, if any, pests and diseases cause problems to the successful cultivation of this attractive leucospermum. It is fairly frost hardy once established.

PROPAGATION:
Easily grown from seed, and shows little variation in the progeny produced. It can also be grown from cuttings, which may be taken from semi-hardwood in early autumn.

USES:
A larger shrub suitable for gardens where space is not at a premium. The flowers are particularly striking, both on the bush and when picked for floral arrangements. The large leaves are an attractive feature of the bush when flowers are not present, and it should warrant a place in most larger gardens. The flowers will not be suitable for commercial purposes as the large heads, unfortunately, are easily broken off the stems.

PLATE 76: *Leucospermum glabrum* (variants).

LEUCOSPERMUM LINEARE

Grows in the mountains of the south-western Cape at altitudes of 300 to 1000 m in gravelly clay soils derived from granite. Rainfall is 1000 mm or more per annum and is concentrated over the winter months. Winters are usually cold and summers hot and dry. Commonly known in South Africa as Narrow Leaf Pincushion.

DESCRIPTION:

Growth habit: generally an erect shrub of neat proportions to 2 m in height. Occasional specimens may be seen that have assumed lower, bushy dimensions of 1.5 m with an equal spread.

Leaves: a distinguishing feature of this species. Particularly long and very narrow at only 2 to 7 mm wide and 50 to 100 mm long. Deep green in colour and often tipped with red.

Flowers: borne on long stems of 400 mm or more. Colour may vary from pale yellow through soft apricot shades to brilliant red. The flowers are neat and very symmetrical in their shape and are normally held stiffly erect. The main period of flowering is from September to December. Unlike many other *Leucospermum* species which may have more than one flower per stem, *L. lineare* normally produces a solitary head which may measure up to 100 mm in diameter.

Variants: as mentioned above, flower colours vary from soft yellow through the apricot shadings to deep reds with intermediate colours. Growth may vary slightly also. One of the most interesting variants is a hybrid 'Flame', of New Zealand origins. Noted for its brilliant orange-red colouring.

CULTIVATION:

L. lineare has proved adaptable and grows easily in most well-drained, acid soils. It withstands dry conditions and quite heavy frosts once it has become established. A good pruning after flowering will keep the bush thick and compact. No particular pests or diseases have been observed to affect this species under cultivation apart from the occasional attack from leaf-roller caterpillars. These can be controlled through the use of a good insecticide.

PROPAGATION:

Seed can give variable results, progeny produced sometimes showing hybrid influence, as in the case of 'Flame'. Plants showing desirable characteristics should be vegetatively propagated, i.e., by taking cuttings of semi-hardwood material during late summer. It has been found that many leucospermums root more easily from cuttings when 'bottom heat' is used in a glasshouse.

USES:

L. lineare is a particularly attractive leucospermum as the narrow foliage shows the flowers to advantage. As a garden plant it has the merits of often flowering slightly earlier than many other species and its blooms persist on the bush for a lengthy period and are very useful for picking. In arrangements, the flowers are seen to advantage because of the lighter textured foliage, and the vase life is particularly good.

PLATE 77: *Leucospermum lineare* (colour variations).

LEUCOSPERMUM MUIRII

Located in a very restricted area of the southern Cape where it is seen growing in deep, white sands of recent origin. The habitat is quite exposed to coastal winds and is fairly temperate. Most rainfall occurs during winter. Other Proteaceae growing in association with *L. muirii* include *L. praecox* and *L. cuneiforme*, *Protea repens* and *Leucadendron galpinii*.

DESCRIPTION:
Growth habit: an erect, bushy shrub of 1.5 m in height, with a spread of 1 m. It forms a short trunk before the densely branched habit begins.
Leaves: medium, grey-green in colour, 40 to 50 mm long and 10 to 15 mm wide. A varying number of teeth can be seen at the leaf apex, ranging from three to seven, and each is tipped with a small, red, rounded peak. Younger foliage is often quite hairy, but this is soon lost with maturity.
Flowers: smaller than most of the better known *Leucospermum* species, each inflorescence being no more than 20 to 30 mm in diameter. They are clustered at the ends of the stems in groups of two, three or four. Colour is a light yellow which gradually changes to a predominantly orange shading, this in turn becoming a deep orange-red with age. All three colours may be seen on the bush at the same time. Stem length is from 200 to 300 mm. The main flowering time is from spring to mid summer.

CULTIVATION:
Generally regarded as a simple species to grow in most well-drained, acid soils, it nevertheless seems to respond to lighter, sandy soil types. It has proven to be quite frost hardy, reliable and long-lived under cultivation. Aside from the occasional caterpillar damage to foliage or stems, no particular pests or diseases have shown to be problems.

PROPAGATION:
Can be raised from seed which shows a little variation. Most plants grown by nurserymen are propagated from cuttings taken in late summer.

USES:
A showy, smaller species for a variety of garden positions. The blooms have proven suitable for smaller, posy bowl arrangements and similar applications, and have a good vase life.

LEUCOSPERMUM PRAECOX

As for *L. muirii*, being found in a similar location and growing in the same sandy soil type.

DESCRIPTION:
Growth habit: an erect rounded shrub of 2 to 3 m in height, rather like a larger version of *L. muirii*.
Leaves: 35 to 60 mm long, and 15 to 30 mm wide. Notably toothed at the apex. Medium green in colour, pubescent when young, glabrous when mature.
Flowers: very similar in overall appearance to *L. muirii*, but considerably larger at up to 60 mm in diameter. Up to four being produced in the terminal clusters. They go through the same colour changes as *L. muirii* and are produced at a similar time of year. Stem length is up to 350 mm.

CULTIVATION:
Again, like *L. muirii*, this species is adaptable to most well-drained, acid soils, but prefers lighter, sandy conditions.

USES:
Ornamental garden shrub for a variety of applications and may have some floral potential.

PLATE 78: *Leucospermum muirii* (upper), *Leucospermum praecox* (lower).

LEUCOSPERMUM PROSTRATUM

This species is found growing in a restricted area of the south-western Cape. Usually seen on sandy, coastal flats it also may be found in hilly country, but soils are always loose, sandy, markedly acid (often less than pH 4) and are very poor. Rainfall is mainly over winter and in the range of 635–1000 mm. Frosts are minimal.

DESCRIPTION:
Growth habit: forming a prostrate mat covering an area of 1 to 2 m wide, this species never exceeds more than a few centimetres in height, and has long, trailing branches which bear the flowers.
Leaves: soft, grey-green in colour and slightly pubescent (hairy), particularly on young growth. Narrow, small by leucospermum standards, 20 to 40 mm long, and only 2 to 6 mm wide.
Flowers: small, round, 20 to 30 mm in diameter, carried in profusion over spring and summer, but some flowers may occur at other times of the year. At first, as they open, the flowers are a light yellow colour, but as they develop they become soft orange, then gradually age to a dark orange-red. As all stages of flower maturity occur together on a bush, a spectacular multi-colour effect results. Few leucospermums have scented flowers, but this is one, having a noticeably sweet, citrus-like perfume.

CULTIVATION:
L. prostratum requires a very light, sandy soil that is more acid than normal. In South Africa it has a reputation for being susceptible to soil-borne fungous diseases when grown in heavier soils. Fertilisers should be completely avoided. Full sun and protection from heavier frosts would seem essential for successful cultivation.

PROPAGATION:
From either seed or cuttings taken from soft to semi-hardwood in late summer.

USES:
An ideal plant for hanging or cascading over banks and walls where its spectacular floral display can be seen to best advantage. It may also make an unusual and intriguing carpet or ground cover plant in the foreground of shrub borders in suitable areas. Useful for various floral art applications. It is sometimes marketed under the commercial name 'Ground Fire'.

PLATE 79: *Leucospermum prostratum.*

LEUCOSPERMUM REFLEXUM

This well-known species occurs in a very restricted area of the Cape at higher altitudes of 1000–2000 m. While the area is arid with low winter rainfall, *L. reflexum* is usually found in moist situations where seepage occurs or near natural water sources such as stream banks. Commonly known as Rocket Pincushion or Skyrocket Leucospermum.

DESCRIPTION:
Growth habit: the largest-growing leucospermum, *L. reflexum* can attain a height of 4 m with an equal spread. Dense and very bushy in habit, branches spread out virtually from ground level.
Leaves: small by comparison to many other leucospermums, 50 mm or less in length, 10 mm in width, covered with very fine, grey hairs, giving an overall grey-green appearance. *L. reflexum* var. 'luteum' has a lighter colouring to the foliage.
Flowers: commence from September, peak during October and November in most districts and continue into January in Southern Hemisphere gardens. Carried in massed profusion, up to 1000 blooms being recorded from a single mature bush. The prominent terminal flowers are an orange-red colour, and as the styles reflex and bend downwards a more crimson-red colouring is evident. With the recently discovered variety 'luteum' the colouring is a clear, light yellow. Size of flowerheads is generally 100 mm in diameter, and the overall length of the bloom with styles reflexed is up to 150 mm. Stem length is up to 500 mm on average.
Variants: some colour variation can occur with plants grown from seed, and differences in flowering times have also been observed. The most significant variant is *L. reflexum* var. 'luteum', first located by South African botanists in 1973, with reported sightings since 1970. A particularly clear yellow colour, it is very worthy of cultivation. Some nurseries are marketing this variety under the name of 'Cape Gold'.

CULTIVATION:
As with most proteaceous plants, *L. reflexum* requires an acid, free-draining soil in an open, sunny position. Often the largest and healthiest old specimens can be seen growing in coastal areas where soil is very sandy. Frost can cause some damage to young growth, and hard frosts, in excess of $-7\,^{\circ}$C, may prove fatal, particularly to younger bushes. In such situations some form of protection will obviously be required. This species is generally wind hardy, and little damage is known to occur in exposed situations — although some support in the form of a sturdy stake can be advantageous while young plants are establishing.

L. reflexum may be subject to the occasional caterpillar attack, including one that bores into the stems near the ends and causes young tip growth to collapse. A tip fungous disease can sometimes cause die-back of young growth. In both instances proprietary brands of spray chemicals are available to control the problem.

PROPAGATION:
Generally simple to propagate from seed, which produces plants that are usually true to type. Cuttings taken from wood which is on the softer side of the semi-hardwood stage in late summer are also satisfactory.

USES:
A bold, larger shrub for landscape uses, as a specimen, for background shelter or for broad cover on banks. Suitable also for picking purposes, the spectacular 'skyrocket' appearance of the long-stemmed blooms adding appeal to flower arrangements. The lasting quality of the blooms when picked is good. Ideally they should be picked at the stage when the first styles begin to reflex.

PLATE 80: *Leucospermum reflexum*,
Leucospermum reflexum var. 'luteum' (upper right).

LEUCOSPERMUM TOTTUM

Found in more mountainous areas of South Africa's Cape at altitudes ranging from 300–2000 m above sea level. Soils are exclusively of sandstone types while the climate is variable, from hot, north-facing slopes where rainfall is low, to cooler south-facing slopes with higher rainfall.

DESCRIPTION:
Growth habit: a much branched, spreading shrub, often to a greater width than height, reaching from 1 to 2 m wide, but rarely more than 1.5 m high.
Leaves: Deep green, not as noticeably 'toothed' or 'notched' at the tips as many other *Leucospermum* species, sometimes not at all. Usual length is in the order of 40 to 60 mm.
Flowers: the illustration depicts a solitary species on the upper left, while the other samples are of a *L. tottum* x *L. cordifolium* cultivar *L.* 'Fantasy' which is being quite widely cultivated in several countries, both as a garden plant and a commercial cut-flower crop. The species has prominent styles. Both flower over spring and early summer. The hybrid form has long flowering stems while the species has short stems to little more than 400 mm.

CULTIVATION:
Well-drained, lighter soils in open, sunny positions are preferred by this leucospermum. As with most species, caterpillars can cause quite severe damage to both flowering stems near the tips of younger growth and to the foliage. Spraying with a specific insecticide may be necessary, more especially during the spring months.

PROPAGATION:
Propagation from seed is likely to produce plants that may be variable. It is therefore advisable to propagate vegetatively from semi-hardwood cuttings.

USES:
An attractive shrub for the home garden, its broad yet compact shape making it suitable for either general use or specifically for bank cover. The hybrid form *L.* 'Fantasy' has a broader growth habit to 2.5 m, making it even more suitable for the latter usage. 'Fantasy' has proven to be an excellent cut flower, while the species, having both a shorter stem and vase life is not so suitable.

PLATE 81: *Leucospermum tottum* (upper left),
Leucospermum 'Fantasy' (right).

LEUCOSPERMUM TRUNCATULUM

Formerly known as *L. buxifolium*, this species is widely distributed throughout the southern Cape in South Africa. It grows almost exclusively on sandy, acid soils which are derived from Table Mountain sandstone. It is found at altitudes from sea level to 400 m where climates are generally mild. Average rainfall varies from 700–1100 mm and occurs mainly during winter. This species is normally found growing in dense stands rather than as solitary specimens.

DESCRIPTION:
Growth habit: variable, from 1 m to 2 m in height and either sparsely or densely branched, but always erect. Soil and climate appear to dictate growth habits. Under cultivation, in soils that are higher in nutrients and with more shelter than the natural habitat, this species forms an attractive, neat, small shrub.
Leaves: small for a leucospermum, being no more than 10 to 25 mm long and 5 to 10 mm wide. Covered with a soft, silvery pubescence, the leaves are an attractive feature of the plant.
Flowers: much smaller than most other *Leucospermum* species, delicately coloured in soft, pastel shades of cream to light orange, each bloom of the dense, clustered inflorescence deepening with age to become a dark orange shade. Individual flowers measure only 20 mm in diameter, and there may be up to six per cluster on the longish stems of up to 300 mm. The main flowering period is over spring.
Variants: although no variations have been observed there is the possibility of hybridisation being carried out with other smaller flowered species which may result in interesting progeny.

CULTIVATION:
Although this species' natural habitat would tend to the assumption that it would grow most satisfactorily in acid, sandy soils, gardeners have found that *L. truncatulum* is easily grown in most well-drained soils that need not necessarily be particularly acid. It appears to be moderately frost hardy and is not very specific in its requirements. No particular pests or diseases have shown to be a problem, and one of the only recommendations to be made with this species is that a pruning back of flowering stems when blooms have finished will encourage bushy new growth and keep the bushes more compact in habit.

PROPAGATION:
A relatively simple leucospermum to raise from either seed or cuttings. The latter should be taken from material which is a little on the soft side of semi-hardwood, and is usually ready in late summer.

USES:
An unusual leucospermum which will undoubtedly find its way into enthusiasts' collections, *L. truncatulum* will have uses for smaller arrangements, in particular posy-bowl work, the smaller, delicately coloured flowers having considerable appeal. Vase life is good. As a smaller shrub of distinctive appearance it will have a place in shrub borders of the home garden, the neat, felted foliage giving an appearance that is quite unique.

PLATE 82: *Leucospermum truncatulum.*

LEUCOSPERMUM VESTITUM

Found growing in inland areas of the south-western Cape in South Africa, this species inhabits varying altitudes of 60–1300 m in both the mountains and on the flats. Soils are well drained and of sandy composition. Most areas in which *L. vestitum* grows are notably arid, receiving as little as 250 mm rainfall per annum, and this mainly in winter. Summers are hot as well as dry, while winters are frequently very cold, especially at the higher altitudes. Formerly known as *L. incisum*.

DESCRIPTION:
Growth habit: usually seen as an erect bush to 2.5 m in height, but some forms in nature may assume lower, sprawling habits.
Leaves: distinctively toothed or 'notched' at the apex, deep green in colour 50 to 70 mm long and 20 to 30 mm wide. Slightly pubescent in younger growth, this is shed at maturity, the leaves soon becoming quite glabrous.
Flowers: generally similar to *L. cordifolium* in appearance, but inclined to be more flattened at the top. The styles are usually either a dark yellow or light orange, while the rolled back perianths or 'ribbons' around the central cone are either purple or dark red and darken noticeably with age. The flowers may be up to 90 mm in diameter and are carried on the bush over early summer. Stem length is quite good at 400 mm or more.
Variants: variable when grown from seed, *L. vestitum* has given rise to a number of named variants, some of which are very deep orange-red, while others are a lighter orange. A brick-orange shading is more typical.

CULTIVATION:
Regarded as a relatively undemanding species under cultivation, *L. vestitum* should prove useful in very dry situations and it is able to withstand cold conditions once established. Most well-drained, lighter soils should suit this species provided they are acid in character and no fertiliser has been added. No particular diseases have been observed to date with this species, and apart from occasional caterpillar damage to foliage and flowering stems, few pests seem to trouble it.

PROPAGATION:
As there are several distinct variants being marketed, it is desirable that propagation should be from cuttings to perpetuate the recognised features of the parent plant. The material is ideally taken in late summer before the wood becomes too firm. Seed will undoubtedly give variations, and these could give rise to some interesting hybrid forms.

USES:
Becoming recognised as a suitable species for commercial picking purposes, the cut flowers have good vase life and pack and travel well. As a garden plant, the compact growth habit and extended flowering period should make *L. vestitum* popular. In addition, it is a reliable and long-lived species.

PLATE 83: *Leucospermum vestitum.*

Zoë C

MIMETES CUCULLATUS

The genus *Mimetes* is found exclusively in South Africa and there are only 11 species, two of which are in danger of becoming extinct. *M. cucullatus* (formerly *M. lyrigera*) is the best known species and is sometimes commonly referred to as Red Bottlebrush in South Africa. It is widely distributed throughout the southern Cape Province in well-drained, sandy or gravelly soils that are fairly acid and high in humus content. In many areas it is found close to the sea in sandy soils. The climate tends to be humid and is not subject to extremes of temperature.

DESCRIPTION:
Growth habit: often sprawling and untidy in the first few years of growth, it does eventually form a neat, rounded bush of up to 1.5 m in height.
Leaves: medium to deep green, carried densely and in neat formation on the stems, 15 to 20 mm wide, 60 to 80 mm long, delicately toothed at the ends with a light red colouration to these tips. For much of the year the leaves towards the ends of the flowering branches are coloured bright red.
Flowers: particularly distinctive, instantly setting the mimetes apart from other Proteaceae. The flowers are composed of silvery, fluffy stamens and projecting red styles and are effectively enclosed and hooded by the brilliant, scarlet red bracts which shade to a yellow colouring at their base. Flowers may be found on the bushes at almost any time of year. Stem length is normally 20 to 30 cm.
Variants: some variation may occur with plants grown from seed. This is usually only in the shade of the red which may be deeper in some cases, tending towards a crimson colour, or lighter, to a more yellow toning.

CULTIVATION:
Most average, well-drained soils that are acid and have a reasonable humus content should be suitable for growing this species. It is able to withstand light to medium frosts provided they are not too prolonged. Salt wind in coastal gardens should pose no problem to *M. cucullatus* as it is a coastal species. It can be pruned quite heavily once established as it has a vigorous root stock from which the bush springs away after veld fires.

PROPAGATION:
Comparatively easy to propagate from either seed or cuttings of semi-hardwood taken in spring or autumn.

USES:
A very colourful shrub for a variety of landscape purposes. Some red bract or leaf colour is present almost throughout the year. As a cut flower it is long lasting and quite unique in its appearance. As it is usually long-lived and reliable once established, it should be a popular garden subject.

PLATE 84: *Mimetes cucullatus.*

Zoë C

PARANOMUS REFLEXUS

Paranomus is an unusual and little known genus of 18 species, *P. reflexus* being the best known and most easily cultivated. The natural habitat is a restricted area of the Van Stadens Mountains near Port Elizabeth in South Africa where only a limited rainfall occurs, but this is evenly distributed. Soils are generally well drained, poor and rocky.

DESCRIPTION:
Growth habit: dense, bushy, branching from ground level, and forming a large bush of 2.5 m in height and nearly as wide.
Leaves: finely divided, not unlike those of a serruria in general appearance. Light to medium green, pointed, with red tips. Overall length is 70 to 80 mm and they are densely carried on the long stems. An unusual and distinctive characteristic is the manner in which the upper half of the flowering stems dramatically change their foliage to become broad, conventionally-shaped leaves, undivided and 20 mm wide.
Flowers: carried in profusion over the winter months with striking buds becoming evident from autumn. Produced on long stems of 500 to 600 mm, the blooms are terminal, 'bottle-brush' like in appearance, and are a subtle creamy-green to buff colour. The size is normally up to 120 mm in length and 80 mm across when fully open.

CULTIVATION:
An easily grown shrub in average, well-drained, acid soils, but needing an open, sunny position to give good flowering performance. As with all proteaceous plants, fertilisers should be avoided. It is drought tolerant once established and will withstand some frost, although young plants should be protected. Pest or disease problems are rarely encountered, and most gardeners should find this species simple to cultivate.

PROPAGATION:
From seed which may be harvested over summer, or from semi-hardwood cuttings which should be taken in late spring or early autumn.

USES:
An unusual, bold-effect shrub for larger gardens. Popular for floral art purposes, the tall flower spikes lending unusual, subtle colour to arrangements. Vase life is good, the blooms lasting for 2 or 3 weeks, especially in cooler weather.

PLATE 85: *Paranomus reflexus.*

Zoë C

PROTEA AMPLEXICAULIS

A distinctive species which is widely distributed in the central mountain ranges of the south-western Cape in South Africa. It is usually found growing in soils derived from sandstone on dry, sunny slopes. Its habitat is frequently subjected to severely cold winters.

DESCRIPTION:
Growth habit: forms a dense, sprawling mat 300 to 400 mm high by 1.5 m wide. Seen to best advantage cascading over banks or walls.
Leaves: particularly distinctive foliage that is reminiscent of some of the *Eucalyptus* species. Heart-shaped, stem-clasping and a striking blue-grey colour against which the pink young growth contrasts vividly making the leaves the dominant feature of *P. amplexicaulis*. In many districts, the colder the winter the more colourful the foliage becomes. May be up to 80 mm long and 50 mm wide.
Flowers: concealed among the branches, particularly on the older wood, the flowers are rarely seen unless the branches are parted or lifted up. Borne during winter in most districts, later in colder inland areas, the flowers are very short stemmed, bowl shaped, up to 80 mm in diameter, and velvety brownish-purple on the outer surface of the bracts, nearly white on the inner surfaces. The central flower mass is basically pale, creamy yellow with red tips. The overall effect of the flower is most unusual and quite striking.

CULTIVATION:
Suitable for growing in lighter soils that afford the good drainage required. Poorer, acid soils appear to be preferred. Seen to best advantage when planted in a position to spill over a wall or down a steeper bank. Withstands heavy frosts. Areas experiencing warm, humid conditions may find *P. amplexicaulis* subject to some fungous dieback of the foliage. This can be controlled through the use of a systemic or copper fungicide. Few other problems are likely to be encountered.

PROPAGATION:
Propagation from cuttings taken from semi-hardwood in spring or autumn is satisfactory. Seedling plants exhibit little or no variation, generally being true to type.

USES:
A charming and unusual cover plant for banks or walls, the foliage and habit creating a distinctive landscaping feature. Flowers, while not a feature of obvious merit, are nevertheless sometimes picked for floating in bowls or for wiring for posy-bowl work. An unusual protea whose potential is probably greatly underestimated.

PLATE 86: *Protea amplexicaulis*.

Zoë C

PROTEA ARISTATA

Found growing in nature in a restricted mountainous area at altitudes of 750–2000 m. Rocky, inhospitable situations where little else grows are favoured. The climate is generally arid and cold in winter. In their natural habitat, specimens judged to be at least 50 years old have been observed. Commonly known in South Africa as Ladismith Protea or more appropriately Small Pine Sugarbush.

DESCRIPTION:
Growth habit: slow-growing to form a neat, erect bush of some 1.5 m after 4 to 5 years. Very old, mature bushes may reach to 2 m. Regarded as one of the longest lived proteas, and one that always retains a neat, tidy habit, even in old age.
Leaves: notably different from most other species of *Protea*, being needle-like, 70 to 110 mm long, but only 2 to 3 mm wide. Medium, glaucous-green in colour, borne in quite dense formation on the long stems.
Flowers: of a typical, narrow cup-shape, up to 140 mm long and 100 mm wide. Brilliantly coloured in crimson, with a silvery 'bloom' creating a soft effect on the bracts, the flowers are usually produced over summer. Unfortunately when picked, the foliage gives off an odour that is offensive to some.
Variants: no particular variants are recorded, apart from mention in Dr. J. P. Rourke's book *The Proteas of Southern Africa* of a natural hybrid between *P. aristata* and *P. nitida*.

CULTIVATION:
Well-drained soils, acid in character, that are either rocky or sandy should suit. Dry and frosty conditions, being similar to the conditions in the natural environment should pose no problems. On the other hand, warm, humid conditions may cause some fungous problems that could necessitate spraying. Foliage does not appear to be affected by any pests. *P. aristata* should prove to be one protea that is ideally suited to colder inland areas, although young plants may need protection from frost while becoming established.

PROPAGATION:
Fresh seed germinates readily and as a rule is true to type. Cuttings from semi-hardwood taken in late spring or autumn are fairly easy to strike and will usually give plants that flower earlier than seedlings.

USES:
A particularly desirable protea for many situations. Neat, erect habit, long life, especially beautiful flowers and unusual foliage all contribute towards *P. aristata* being suitable for a variety of applications. Its frost-tolerance will no doubt add to its appeal for gardeners in colder districts.

PLATE 87: *Protea aristata*.

PROTEA AUREA

Better known by its more appropriate former name, *P. longiflora*, this widely distributed species is found growing in the coastal mountains of the southern Cape, predominantly in cooler, moist situations on south-facing slopes. Soils are well drained, generally poor and vary from acid to slightly alkaline.

DESCRIPTION:
Growth habit: fast-growing to form a dense, erect bush 3 m in height and often 2 m or more across. These dimensions are usually reached within 3 to 4 years.
Leaves: broadly oval in shape, comparatively short, measuring 25 to 40 mm in width and 50 to 80 mm long. Deep, greyish-green in colour and may have a slight, velvety pubescence, particularly on younger growth.
Flowers: variable in colour, from a creamy white through intermediate pink shadings to deep rosy-red. They are borne on quite long stems of 400 to 500 mm during late summer to mid winter as a rule, but some flowers may appear at other times of the year. Shape at the bud stage is particularly long and narrow, up to 130 mm long and only 20 mm in diameter, but opens to form a broad, 'shuttlecock'-shaped flower up to 120 mm.
Variants: apart from the colour forms mentioned, the only other variant recognised is a subspecies, *P. aurea* subspecies *potbergensis*. This is a striking plant with larger, creamy-green flowers and broader, bright green foliage.

CULTIVATION:
Regarded as a very adaptable and easily grown protea. Most well-drained soils are suitable, and it is notable for withstanding winds, frosts and poor soils very well. It has few if any pest or disease problems of significance and is easily grown in most districts.

PROPAGATION:
May be grown from seed which, while true to the parent plants' basic characteristics in so far as growth habit goes, may be variable in flower colouring. Selected colour forms should be propagated from cuttings taken from semi-hardwood in later spring or autumn.

USES:
The vigorous growth, large dimensions and long flowering period combine to make *P. aurea* a particularly useful protea for colourful background or screening purposes in the garden. The flowers, while most attractive at the mature bud stage and when first open, tend to collapse in the centre fairly quickly after picking, limiting its usefulness for cut-flower purposes. In spite of this, it has proved to be quite popular in floral work.

PLATE 88: *Protea aurea*, typical colour forms.

Zoë C

PROTEA CANALICULATA

Better known by its former name, *P. harmeri*, this charming, smaller-flowered species is found in mountainous regions of the south-western Cape, where summers are very hot and dry and winters are cold with the occasional snowfall being experienced. Soils are particularly rocky, well drained, poor and acid as a rule. Favoured sites are normally very exposed, receiving maximum sun and some wind.

DESCRIPTION:
Growth habit: normally a fairly erect, small shrub to 1 m or slightly more, but sometimes may bush out to form a shrubby habit.
Leaves: variable, some forms have needle-like leaves as little as 2 mm wide and 50 to 60 mm long, while others have slightly broader foliage measuring up to 7 mm wide and 70 or 80 mm in length. Colour is also variable, ranging from deep green through to glaucous, blue-green. It is usual for leaves to show a slight curl or twist.
Flowers: again variable, ranging from deep reds to softer pinkish-rose shades, with the backs of the bracts exhibiting silver markings. Broadly cup-shaped, the bracts open wide to reveal the cream and bronze central flower. Size is 40 to 50 mm in diameter. Normal period for flowering is from autumn through to mid winter. Stem length is usually from 100 to 200 mm.
Variants: as already mentioned, flower colour, leaf size, shape and colouring all are variable.

CULTIVATION:
Regarded as an adaptable species, thriving under both moist and dry climatic conditions. *P. canaliculata* is unfortunately subject to some foliage blackening caused by a fungous disease which is more evident in humid climates than in drier inland areas. It can be controlled through spraying with a copper or systemic fungicide. Most well-drained, acid soils in open, sunny positions seem to suit this species. Pests do not normally pose any problem.

PROPAGATION:
Easily propagated from seed which does not as a rule show much variation from the parent plants. Cuttings may be taken from semi-hardwood in late spring or autumn.

USES:
A charming smaller species for the garden, its compact size fitting into many landscape situations with ease. Useful for floral art purposes, but doubtful for commercial picking due to the slightly unpleasant odour of the foliage. A beautiful smaller protea for the collector or enthusiast.

PLATE 89: *Protea canaliculata.*

PROTEA 'CLARK'S RED'

A significant New Zealand hybrid selected by nurseryman Jack Clark in 1948. He observed the native nectar-seeking tui moving between *P. repens* and *P. aurea* and anticipating a natural hybrid, saved the seed. One seedling was different enough to isolate and this has become known as 'Clark's Red', a very popular plant with New Zealand gardeners, and becoming available in Australia.

DESCRIPTION:
Growth habit: neat, densely erect shape to maximum of 3 m high and 2 m wide. Fast-growing reaching 2 m in 3 years.
Leaves: considerably broader than the *P. repens* parent and more leathery. Narrower than the *P. aurea* parent. Usually 80 mm long and up to 20 mm wide. Medium to dark green and of firm texture. Young foliage is sometimes seen to have a pale red margin.
Flowers: similar in overall appearance to the *P. repens* parent, although slimmer in dimension. Particularly rich red colour, lightening slightly towards the base of the bracts. Slight silver cast overlaying the red base tonings. The tips of the perianth limbs which encase the styles are softly pubescent and a striking silver, contrasting vividly against the red of the remainder of the flower which is 100 mm long and opens to form a 'shuttlecock' shape 80 mm wide at the tips. Blooms virtually throughout the year, but particularly freely over late summer and autumn.

CULTIVATION:
As with so many of the hybrid proteas, 'Clark's Red' is comparatively undemanding in its requirements. It has inherited the *P. repens* parent's tolerance of more alkaline soils, but is generally happiest, as with all proteas, in acid, well-drained soils, and an open, sunny situation. It has proven to be particularly frost hardy.

P. 'Clark's Red' has unfortunately inherited one undesirable characteristic from the *P. repens* parent. It is subject to a fungous disease that causes blackening of the young leaf tips. At times this can be disfiguring, but never appears to be fatal once the bush is established. Systemic fungicides can be applied to control the problem. In spite of this genetic problem, *P.* 'Clark's Red' is a handsome and desirable garden plant.

PROPAGATION:
By semi-hardwood cuttings only, taken in late spring or autumn.

USES:
A showy garden shrub, always retaining a neat appearance. The remarkably extended blooming period gives some colour at almost any time during the year. If picked just as the buds soften and begin to show the soft silvery centres, vase life can be up to 2 weeks, although the leaves may blacken and need to be stripped off during that time.

PLATE 90: *Protea* 'Clark's Red', (lower flowers and bud),
Protea repens (centre left),
Protea aurea (top).
Bird: the New Zealand tui *(Prosthemadera novaeseelandiae).*

Zoë C

PROTEA COMPACTA

A popular, local market cut flower in South Africa where it is known as the Bot River Protea. This species' natural habitat is a narrow region of the south-western Cape where it is found at low altitudes of under 100 m and never far from the sea. Soils are naturally poor, sandy and acid in character. *P. compacta* normally grows in dense stands where bushes tend to support each other.

DESCRIPTION:
Growth habit: naturally an erect, slender bush, it may tend to be bushier or sprawling in richer soils. Maximum height in nature is 3.5 m, but most examples of *P. compacta* found in gardens are rarely more than 2 to 2.5 m. It is recognised as being comparatively short-lived, 10 years being the usual life span, although the odd cultivated specimen has achieved 20 years.
Leaves: quite distinctive, 100 mm or more in length and 50 mm wide, soft and hairy while young, becoming more leathery and glabrous with age. Always neat in appearance, and carried quite closely to the stems of the bush.
Flowers: these are a striking clear pink colour, the outer bracts gradually fading to a pale cream towards the base of the flower. Carried on long, straight stems of 400 to 500 mm as a rule. Winter is the usual flowering period, but occasional flowers may be seen at other times of the year. They are 120 mm or more in length and up to 90 mm in diameter.
Variants: when grown from seed, some variations occur. The closeness of the flower bracts and the intensity of the pink colouration, as well as the size of the flower, are all variable, but still remain instantly recognisable as *P. compacta*. In the natural habitat, a white version is sometimes encountered, although it is considerably less common than the usual pink.

CULTIVATION:
Lighter, well-drained, acid soils are required for successful cultivation. *P. compacta* appears to be tolerant of light to medium frosts and does not appear to be adversely affected by salt winds when grown nearer the coast. It is subject to some leaf diseases which cause unsightly blackening of young foliage in particular. This can be controlled through the use of systemic or copper fungicides used sparingly. Insect pests do not normally present a problem.

PROPAGATION:
Fresh seed germinates readily and is usually fairly true to type (see under variants, above). Cuttings taken from semi-hardwood in late spring or autumn are a satisfactory method of propagation also.

USES:
Although having the disadvantage of lasting no more than 10 years or so in the garden, *P. compacta* is nevertheless a very lovely protea for winter display, the clear pink flowerheads showing up well. For floral art purposes it is particularly popular, the neat shape and distinctive colouring adding to its appeal. Foliage of cut stems may tend to blacken after a few days, but this does not seem to diminish its popularity. For garden display purposes the upright habit of *P. compacta* means it can often be accommodated among other shrubs where space precludes the use of many of the bushier protea species.

PLATE 91: *Protea compacta.*
Protea compacta, white form (top).

Zoë C

PROTEA CORONATA

Found growing in scattered populations through the south-western and southern Cape. Occurs mainly on heavy clay soils in areas of high rainfall. Grows naturally in dense stands rather than as individuals. Commonly known both in South Africa and in other countries where it is grown as the Apple Green Protea. Botanically previously known as both *P. incompta* and *P. macrocephala*.

DESCRIPTION:
Growth habit: usually erect and bushy, normally reaching a maximum height of 2.5 m, old specimens may reach 3.5 m and assume a rather gnarled appearance. At its best after 2 or 3 years when it may be seen as a neat, dense bush of up to 2 m.
Leaves: a distinguishing feature. Most forms have silvery-green, pubescent foliage, although some have bright green leaves lacking pubescence or 'hairiness'. Length of the lanceolate leaves is up to 90 mm and they are 8 to 30 mm wide.
Flowers: characteristic 'Granny Smith' apple-green bracts clasping woolly, white or pink 'button' centres of perianth awns. Size of flowerhead is variable, but on average 100 mm long and 50 to 60 mm in diameter. The main flowering time is during winter, with some blooms being produced until late spring.
Variants: three main variations are found among plants offered by nurserymen. One is a narrower flowered type with a pure white button centre, another similar one has quite a bright pink central button, while a further variant produces larger blooms with a dark red ring encircling the white centre.

CULTIVATION:
Generally an easy protea to grow in most average garden situations. Tolerant of a wide range of soil types provided drainage is adequate. *P. coronata* appears to be more tolerant of high phosphorus and calcium levels than many other proteas. In shaded or very sheltered situations it can be afflicted by certain fungous diseases which may cause blackening of the foliage. This can be controlled with a systemic or copper fungicide. The only insects to affect this species with any degree of severity are thrips, which can be controlled with a good insecticide.

Occasionally during flowering, the weight of the blooms may tend to drag lower branches down. These branches should be pruned back, either at flowering or immediately afterwards, to maintain a good shape to the bush. A species which is usually frost-tender, more than -2 °C causing damage.

PROPAGATION:
By either semi-hardwood cuttings or from seed. Seedlings grow quickly and produce flowers in the second or third year. To ensure particular flower form or colour, plants should be cutting-grown, as seedlings may be variable.

USES:
Provides a good focal plant in any garden or in a collection with its distinctive foliage. *P. coronata* is appreciated by many floral artists for the bright green colouring of the blooms. A colour not repeated in such vivid tonings in any other protea, it provides good contrast with other colours in arrangements.

PLATE 92: *Protea coronata* (three main variants).

PROTEA CYNAROIDES

Well known internationally as the King Protea and since 1976 the official national flower of South Africa. Grown in several other countries as a garden plant for 60 years or more. One of the most widely distributed Cape proteas, *P. cynaroides* in its varying forms is found over a large area of the southern Cape, from close to the sea to the tops of mountain ranges, where at 1500 m above sea level it may often be covered by winter snow. It is generally found on well-drained soils in areas that receive 400 mm or more rain per annum. Soils may be as acid as pH 3.5.

DESCRIPTION:
Growth habit: variable growth habits are characteristic of *P. cynaroides*, and in South Africa many forms are recognised, ranging from low-growing dwarf variants to taller bushes of at least 2 m high.
Leaves: varying greatly, from almost circular to oval or even broadly lanceolate. They may measure from 120 to 300 mm in length and 20 to 130 mm in width. Generally medium green, shiny and with a distinct stalk. Often with a noticeable red margin around the leathery, thick leaves, it is the foliage that instantly identifies *P. cynaroides* when not in bloom.
Flowers: as variable as the leaves. Colours range from subtle greenish whites through soft silvery pinks to the deep reds. Size and shape is also variable, some forms having quite small, funnel-shaped blooms, while others have the more well known, wide-open large flowers that may measure up to 300 mm across. Flowering can be at any month of the year depending on the origins of the plant, different regions of the Cape producing distinct races each with their own characteristics. Noted as a cut flower, with cut blooms creating dramatic effect in larger arrangements. Stem length is variable, from 40 to 75 cm.
Variants: Dr. Marie Vogts has identified four main groups of *P. cynaroides*, and from within these at least 80 variants, each characterised by different leaves, flower form, colour and flowering time. Of interest to gardeners and commercial flower growers will be the simple classification of spring, summer, autumn and winter flowering variants. In the natural environment, hybrids from crosses with *P. obtusifolia* and *P. compacta* have both been observed.

CULTIVATION:
Well-drained, lighter soils are most suitable for *P. cynaroides*, and particularly healthy specimens are often seen in sandy, coastal gardens. *P. cynaroides* is one of the few protea species to have a vigorous, fleshy rootstock which enables the bush to virtually regenerate after veld fires have burned off the growth to ground level. Because of this, it is safe to prune the bush back hard when it has reached old age and may have become untidy and sprawling in habit. Cutting back almost to a stump will encourage new shoots in profusion, often resulting in a thicker and bushier plant than was the case previously. In spite of such seemingly harsh treatment, flowers are often produced the year after the pruning, and this technique is practised by commercial growers in South Africa seeking to promote longer flower stems and new, fresh growth. Some chewing insects may cause leaf damage, and even the odd flower bud may show signs of insect damage. This is not common, however, in the home garden and can be overcome to a large extent by spraying with reliable insecticides.

PROPAGATION:
Growing *P. cynaroides* from seed is comparatively simple, and seed taken from specific localities appears to be genetically stable and will exhibit the same characteristics in terms of flowering time as the parent plants. Propagation from cuttings of semi-hardwood is a satisfactory method of perpetuating particular qualities. Often, however, *P. cynaroides* bushes do not produce much propagating material, and this necessitates the building up of a good number of stock bushes from which to procure material; a time-consuming process, and one of the reasons that named cultivars are often difficult to obtain.

USES:
As a spectacular flowering shrub, *P. cynaroides* can have few equals. A selection of the variants in a larger garden can ensure flowering throughout the year. A massed planting of 'Kings' can also be a very dramatic landscape feature when they are in flower. For picking purposes, its uses are limited. So large is the bloom that only one or two at a time can be used and even then only in larger arrangements. Without a 'King', no protea collection can be considered either representative or complete.

PLATE 93: *P. cynaroides* — a large summer-flowering variant.

Zoë C

PROTEA EFFUSA

Found growing in mountainous areas of the western Cape, generally on exposed, barren ridges in rocky terrain at altitudes of 1200–1800 m where winters are cold and frost and snow are commonplace. Formerly known as *P. marlothii.*

DESCRIPTION:
Growth habit: variable, mainly according to the particular locality from which the plants have originated. Local races or 'forms' found in the natural environments range from totally prostrate and mat-forming at no more than 200 mm high in one locality to taller bushes of 1.5 m in another. Spread may be up to 2 or 3 m.
Leaves: as variable as the growth habit of the plants. May be a soft lush green colour and up to 80 mm or more long and 30 mm wide, or a darker, almost glaucous, blue-green shading and smaller in size at 40 to 50 mm long and 10 to 20 mm wide.
Flowers: again variable, but mainly in colouring. Generally shaped like a shallow bowl, 60 to 100 mm in diameter. Colour ranges from greenish-yellow through intermediate shades to deep red on the bracts. The inner flower is normally a honey-gold colouring, deepening towards the apex with dark golden bronze tips. Stem length may be up to 300 mm, flowers may be either semi-pendulous or almost erect, and the main flowering period is from early winter to spring, with the peak normally being reached in mid winter.
Variants: as detailed above, localised 'races' occur in nature with varying growth habits and flower colour. Natural hybrids between *P. effusa* and *P. pityphylla* have been observed in the habitat.

CULTIVATION:
This species should not prove difficult in most home gardens, preferring well-drained, acid soils and withstanding cold winters. The lower, spreading forms are ideally suited to banks, hanging over walls and similar landscape situations. Insect pests do not usually seem to be a problem but in some areas, particularly those with warmer temperatures and higher humidity, some fungous diseases may cause problems with leaf blackening. Systemic or copper fungicides should control this problem when it manifests itself.

PROPAGATION:
Either seed or cuttings of semi-hardwood are satisfactory methods of propagation. Forms that exhibit particularly desirable characteristics are best perpetuated by cuttings, however.

USES:
Highly ornamental subject for bank cover in the case of the spreading form. As a cut flower it is particularly beautiful, and appreciated by both florists and the keen floral artist.

PLATE 94: *Protea effusa.*

PROTEA EXIMIA

Found throughout the more mountainous terrain of the southern Cape, in widely divergent climatic conditions, ranging from mild, moist areas to higher altitudes where extremes of temperature are experienced, including snow in winter. Soil types are generally very well drained, acid and impoverished. Formerly botanically known as *P. latifolia* and commonly referred to in South Africa as the Ray-flowered Protea. Some commercial flower growers refer to it as The Duchess.

DESCRIPTION:
Growth habit: forming an erect bush to 2 m or more within 5 years, and in old age sometimes assuming the proportions of a small, rather gnarled tree to 3 m or more.
Leaves: variable, mainly in colour, ranging from green to glaucous, sometimes with a pink to purple blush particularly during winter. Noticeably broad, oval in shape, with a pronounced red margin, the leaves may measure 100 to 130 mm in length and 40 to 60 mm in width.
Flowers: occur during the late winter and spring as a rule, but occasional blooms may be seen at other times of the year, especially over summer. The widely spaced flower bracts are a bright red shading down to a cream at their base. Through the gaps between the bracts may be seen the central flower mass, which is basically cream in colour with prominent purplish-red tips. Stem length is usually 250 to 400 mm.
Variants: blue-tinted foliage, some variations occur with flower colour, the intensity of the red ranging from almost pink to deep crimson.

CULTIVATION:
Regarded as one of the easiest proteas to grow in a wide range of garden situations. *P. eximia* is remarkably tolerant to variations in soil types from acid to alkaline, and is certainly one of the most frost hardy proteas. It is also generally unaffected by winds in most localities. Very few pests and diseases cause any problems, the odd caterpillar holes in foliage being the only real pest that is seen to affect the shrubs, and this not to any marked extent.

PROPAGATION:
May be propagated from either seed or cuttings, the latter usually being taken from semi-hardwood material. Seed-grown plants are usually fairly true to type, although occasionally some interesting hybrid variations occur.

USES:
A good protea to try in problem areas where other species have failed. Probably one of the most adaptable proteas, it may be seen growing where very few other members of the Proteaceae succeed. Not usually regarded as a suitable species for picking purposes, it nevertheless can be attractive in the vase when first picked before the central flower mass collapses. For garden display, there are few times of the year when some flowers are not in evidence.

PLATE 95: *Protea eximia.*

Zoë C

PROTEA 'FRANCISCAN HYBRID'

A chance hybrid from a seed-grown plant of *P. compacta*. The other parent is thought to be either *P. magnifica* or *P. eximea*. It is included here because it is this type of hybrid which is becoming important in the nursery industry both for home garden and cut-flower purposes.

DESCRIPTION:
Growth habit: large, bushy growth to 3 m and of equal width. Even though the original plant is 30 years old, it exhibits none of the 'leggy' or gnarled look that often characterises old protea bushes. It is still dense and well-clothed with healthy foliage. Apart from picking off spent flowerheads, it has had no special attention.
Leaves: medium to light green, slightly pubescent (hairy) and similar to *P. magnifica* in shape and size, but much softer and less leathery in texture. Approximately 100 to 140 mm long and 30 to 50 mm wide.
Flowers: the most distinctive feature. Similar to a smaller, neat *P. magnifica* bloom in shape and size, but with less of the prominent, bearded tips to the bracts. These bracts are basically a rich pink, shading to a delicate cream at the base, with a distinctive silvery down, which in some lights gives a soft silvery-pink appearance to the blooms. The central flower mass is strongly shaded in a deep wine colour. On average the blooms measure 150 mm in length and 100 mm in diameter. The flowering period is over spring. This may vary slightly from one district to another. Stem length is generally 300 to 400 mm.

CULTIVATION:
Appears to be undemanding in its requirements and is tolerant of heavier clay soils. As with most proteas, however, a well-drained soil that is on the acid side is preferable. It tolerates frost to at least −5 °C, and has not been found to have any particular pest or disease problems.

PROPAGATION:
By cuttings only for the characteristics to be perpetuated. These should be taken from semi-hardwood material in late spring or autumn. Seed would probably give a wide variation in type, and possibly even some throwback to the original parents of the cross.

USES:
Undoubtedly a protea that will prove very popular with both home gardeners and commercial flower growers. As a cut flower its appeal will lie with the dual characteristics of attractive colouring and form plus good vase life and travelling ability.

PLATE 96: *Protea* 'Franciscan Hybrid'.

PROTEA GRANDICEPS

Widely distributed throughout South Africa's southern Cape, this species is usually located in higher, more mountainous areas at altitudes of 1000–1700 m. Soil conditions are generally rocky and poor, and the exposed situations are hot and dry over summer and cold during winter. Commonly referred to as Peach Protea or Princess Protea. The Latin *grandiceps* means large or noble head.

DESCRIPTION:
Growth habit: bushy, rounded, and reaching a height and spread of less than 2 m *P. grandiceps* is fairly slow growing and, as is often the case with slower-growing proteas, is long-lived, 20 years or more being quite commonplace. Particularly old specimens have been known to reach a height of up to 2.5 m with an even greater spread.
Leaves: the large, leathery, grey-green leaves are distinctively margined with red. Generally oval in shape, they may be up to 140 mm long and 80 to 90 mm wide. Carried densely on the stems, the leaves are an ornamental feature of the bush when it is not in flower.
Flowers: the natural flowering period is from mid summer, but at lower altitudes under cultivation flowering time is very variable, spring blooms not being unusual. Colour of the large flowerheads is generally a bright reddish-brown with a prominent brown or white 'beard' around the opening of the bracts. Size of the blooms is usually in the order of 100 to 150 mm long and 80 to 120 mm in diameter.
Variants: some variation in the size of leaves is encountered, as is the colouring and size of blooms; colour can vary from quite light peach or salmon-pink to dark reddish-brown. Beard colour can range from brown to pure white and maroon.

CULTIVATION:
Generally an easy protea to grow in most home gardens, *P. grandiceps* seems adaptable to both good garden soils as well as poorer situations. Heavier frosts appear to be quite well tolerated, although early frosts may burn tender young growth. The main insect pest that can cause problems is white scale, the same well-known pest that causes the unsightly white patches on the leaves of waratahs. Most commonly used control measure is white or 'all-seasons' oil, which is often mixed with other insecticides. Particularly bad infestations may call for stronger measures with specific insecticides being used.

PROPAGATION:
Usually by sowing seed, which normally gives plants true to type. Cuttings are satisfactory, but commercially less practical due to the lack of suitable material in quantity from individual bushes.

USES:
A very ornamental species, *P. grandiceps* is a desirable cut flower with good lasting qualities, an ornamental shrub for the home garden, and appreciated by floral artists. Any protea collection would scarcely be complete without this lovely species.

PLATE 97: *Protea grandiceps*, two typical variants.

PROTEA LAURIFOLIA

Found over an extensive area of the south-western Cape, mainly in the more inland mountains at altitudes of 400–1200 m. Soils are generally poor, well drained, acid and either rocky or sandy in consistency. Summer droughts may be extended, and winters are generally cold with occasional snow falls. Formerly known as *P. marginata*. The Latin *laurifolia* refers to the laurel-like leaf.

DESCRIPTION:
Growth habit: generally erect growing and dense to 3 m under cultivation, but in nature specimens have attained heights of 8 m.
Leaves: generally blue-green in colouring, but dull grey-green in some cultivars. Usually considerably broader than *P. neriifolia*, with which this species is sometimes confused. Width may be up to 55 mm while length can be as much as 170 mm. They are generally thick and leathery, and some cultivars have a red margin to the leaf.
Flowers: in nature the species is usually described as being faded pink or dirty white, and only occasionally as an attractive soft rose. The cultivars have much more distinctive and attractive colouration. The peak flowering time is through the winter, but there is rarely a month when established bushes do not have some blooms in evidence. Flowers are generally long, to 120 mm or more and have a diameter of 70 to 100 mm. They have characteristic black beards to the bract tips, and usually (depending on the cultivar) have purplish tips to the central flower mass. Stem length is good at 300 to 500 mm.
Variants: there are a number of named cultivars of *P. laurifolia*, with a considerable diversity of flower and foliage colour.

CULTIVATION:
Generally among the easier proteas to grow, tolerating a wide range of soil types and climatic conditions. Regarded as one of the more drought-resistant species. As with most species, the main requirements are an acid, free-draining soil, full sun and the avoidance of fertiliser. The only problem in terms of pests or diseases seems to be occasional caterpillar damage to foliage. This can easily be controlled through spraying with a reliable insecticide.

PROPAGATION:
Seed is satisfactory but will usually give variable results with flower colouring varying to a marked degree. Selected forms or cultivars should always be propagated vegetatively by taking cuttings of semi-hardwood in late spring or autumn.

USES:
A long-lived, vigorous garden plant giving excellent background display. Suitable for picking, the vase life being generally good apart from foliage blackening which occurs after a few days with some cultivars more than others. The extended flowering period of most *P. laurifolia* cultivars has made them popular with gardeners. Sometimes used as firewood in South Africa.

PLATE 98: *Protea laurifolia*, variants.

Zoë C

PROTEA LEPIDOCARPODENDRON

Occurs mainly at lower altitudes, often close to the sea, over a fairly restricted area of the south-western Cape. Usually found on heavier clay soils, although it may be seen on lighter sandstone types. In nature it has a preference for reasonably moist soils or areas where regular rainfall is experienced. Climates are generally mild, without very low winter temperatures. Commonly known as either Black-bearded Protea or more simply as The Black Protea. The Latin *lepidocarpodendron* translates to scaly fruit tree.

DESCRIPTION:
Growth habit: an erectly bushy shrub of 2 to 2.5 m in height at maturity. In some situations it can bush out to form a dense plant of 1.5 m in width, and may only reach 1.5 m in height also.
Leaves: similar to *P. neriifolia*, medium to deep green, elongated, 80 to 130 mm and 10 to 20 mm wide. Quite thick and leathery when mature, the leaves are held stiffly erect at only a slight angle to the stem, and may tend to partially obscure the flower.
Flowers: as with the leaves, the flowers are not unlike those of *P. neriifolia* in shape but may be slightly smaller, and have the notably, heavy black beards to the ends of the flower bracts. Size is 120 to 130 mm long and 50 to 60 mm in diameter. Stem length is usually 300 to 400 mm. The main flowering time is winter, but blooms may be seen on the bushes from autumn until the beginning of summer.
Variants: in nature a pink form, also with the prominent black tips, is not uncommon.

CULTIVATION:
A very adaptable protea to varying soil types, *P. lepidocarpodendron* is easily grown in most average garden soils. Heavy frosts in more inland districts are likely to cause damage. It is noted for being resistant to salt winds, which do not seem to cause any foliage damage, and is safely recommended for coastal gardens. Apart from the odd leaf holes caused by caterpillars, this species does not seem to have many pest or disease problems.

PROPAGATION:
Selected forms should be grown from cuttings taken from semi-hardwood in late spring or autumn. Seed is a satisfactory method of propagation but some variation can be expected.

USES:
Suitable shrub for many gardens, having a neat habit and usually proving reliable. Good for coastal areas. Flowers may be picked, although some foliage blackening does occur. The black colouring has proved popular with floral artists seeking the right shadings for dramatic black and white arrangements.

PLATE 99: *Protea lepidocarpodendron.*

Zoë C

PROTEA LONGIFOLIA

Mainly found growing in a limited area of the south-western Cape at lower altitudes, frequently near the coast. Soils are mainly of sandstone origin, sometimes of gravelly clay types, but always free draining and acid. The climate is mild, without extremes of temperature at any season, but notably windy. Commonly known in South Africa as the Long-leaved Protea, which is also the translation of the Latin *longifolia*.

DESCRIPTION:
Growth habit: lower growing, stout, bushy shrubs to no more than 1.5 m as a rule, 1 m being the more usual height.
Leaves: particularly long, from 100 mm up to 200 mm and only 5 to 15 mm wide. Colouring is usually more towards the grey-green tonings. They may often have a distinct twist.
Flowers: variable in colour from delicate greens through orange-browns to orange-reds and pinks. Always with a prominent, pointed, black central peak to the cream flower mass. Size is usually 150 mm long and 70 to 100 mm in diameter. The peak flowering period is during winter, but blooms can be seen on the bushes from autumn through to late spring. Stem length is 250 to 400 mm.
Variants: very variable in flower colour, and to some extent in the size and growth habit of the bushes. *P. longifolia* is one of the most promiscuous proteas, hybridising freely with a large number of other species, including *P. laurifolia, P. compacta, P. repens, P. magnifica* and *P. neriifolia*. Some of these hybrids that have occurred both in nature and under cultivation are particularly attractive.

CULTIVATION:
Easily grown in most well-drained, acid soils, *P. longifolia* will take the odd heavier frost but will suffer damage from sustained heavy frosts. No particular pests or diseases cause problems, the only damage observed being occasional caterpillar holes in the flower bracts.

PROPAGATION:
Plants grown from seed are likely to be very variable, colours being especially so. Hybrids are also frequently noticed with this method of propagation. To ensure particular characteristics of the parent plant, propagation from cuttings is necessary. These will need to be taken from semi-hardwood during either late spring or autumn.

USES:
P. longifolia is well suited to many landscape applications, particularly where a lower, spreading shrub is required for bank cover. It responds well to pruning, especially after flowering, and this encourages new growth which will keep the bush compact longer.

Its use as a cut flower is not likely to be significant, but some forms have proved popular with floral artists who appreciate the unusual colour contrasts exhibited in the blooms.

PLATE 100: *Protea longifolia,* typical variants.

PROTEA MAGNIFICA

Although some gardeners still think of this beautiful protea by its former name, *P. barbigera*, most have now accepted the botanists' revision to *P. magnifica* — a particularly appropriate change. This is a widely distributed species, being found over an extensive area of the south-western Cape. Typically it is found in dry, exposed situations which are hot in summer and cold in winter, snow being quite common. Soils are generally poor, rocky, acid and well drained. Altitudes range from as low as 600 m but more usually are in the region of 1200 to 2700 m. Commonly referred to as the Queen Protea or Bearded Protea.

DESCRIPTION:
Growth habit: in nature a very variable protea, ranging from low, sprawling or prostrate, to rounded, bushy shrubs, or erect, small trees of up to 3 m in height. Most plants of *P. magnifica* sold by nurseries grow into the rounded, bushy type, ranging in height from 1.5 to 2 m as a rule, although occasionally more erect, taller specimens will be encountered.
Leaves: may vary somewhat, but generally are large, leathery, grey-green in colour and measure 100 to 200 mm in length and 30 to 60 mm in width. They are prominently veined, and often have a pronounced red margin.
Flowers: again, these are variable, in size, colour, amount of beard to the tips of the bracts, and also in the time of year they are produced. Size is invariably large, up to 150 or 160 mm in length and up to 150 mm across when fully opened. The white or cream central flower mass is nearly always domed, finishing in a prominent pointed peak. This peak is often black, but may sometimes be brown or without any distinct colouring. The amount of fluffy 'beard' to the surrounding flower bracts is variable, but is usually white, although sometimes can be brown or with a slight touch of black. The predominant colouring of the bracts is usually a soft pink to deeper rose but the creamy-white variant is also seen. In nature, the colour spectrum is wide, ranging from pale greenish-whites through the soft pinks to deep rose and reds. Main flowering time is generally from mid winter to mid summer, but some forms may flower at other times of the year. Stem length is usually from 250 to 400 mm.
Variants: as mentioned above, there are wide-ranging variations in flower form, colour and flowering time. Growth habits are also variable. In nature, numerous hybrids have been observed, and many hybrids are becoming available in the nursery trade.

CULTIVATION:
Generally an easy protea to grow in average, well-drained, acid soils in most areas. Tolerant of frost to a reasonable degree, will take some wind but may exhibit foliage burn from excessive salt winds. Not affected by insect pests to any marked degree, but can be seriously affected by foliage fungous diseases in some instances. This problem can usually be controlled by a conscientious spray programme using copper or systemic fungicides.

PROPAGATION:
Seed gives plants that are normally fairly true to type, but flower colour in particular may be variable from this method. Propagation of selected forms should be carried out by taking cuttings of semi-hardwood in late spring or autumn, this being the only way to ensure that the desired characteristics are perpetuated.

USES:
A particularly handsome and spectacular garden shrub for many situations. Flowers are normally carried over a long period and may be picked, giving a good vase life. This species is suitable for export, the cut blooms travelling well. Floral artists enjoy using the blooms also, dramatic impact being created by the large, bearded flowers.

PLATE 101: *Protea magnifica.*

PROTEA NANA

Formerly well-known as *P. rosaceae*, and commonly referred to as the Mountain Rose, this delightful smaller species is found growing in dense 'stands' at elevations of up to 900 m in a restricted mountainous area of South Africa's south-western Cape Province. Soils vary from moist peat soils to drier sandstone types. In some areas light snow falls are experienced and winter frosts are not uncommon. The Latin *nana* translates to dwarf.

DESCRIPTION:
Growth habit: neat, compact and erect-growing to a maximum height of 1.25 m. Fairly densely branched. May tend to have a slightly sprawling habit for the first season after planting out.
Leaves: a distinguishing feature. Light to medium green, needle-like to 20 to 30 mm long and only 1.5 mm wide. At times may have a slightly reddish cast.
Flowers: buds start forming during late summer and autumn, opening to pendulous, bell-like or 'cup-shaped' flowers from mid winter to late spring. Size can be variable, from 30 to 45 mm in diameter. Colour is generally a deep mahogany or wine-red on the bracts, but may vary through clear red to a reddish-green, the latter being uncommon. The central flower is a buff colour, deepening towards the outer tips.
Variants: flower colour and size are the main variations to occur, as mentioned above. However, a spreading, prostrate form is sometimes encountered that has greenish flowers.

CULTIVATION:
P. nana appears to be one of the more difficult proteas to maintain in gardens with an expected life span of 5 to 10 years. The reason for this is not clear, but it seems apparent that slightly heavier, though still well-drained, soils may be preferred. Some of the best specimens may be seen growing in silt soils: these plants being longer-lived than others.

Frost does not appear to be a problem to this species, this being in accordance with the natural habitat. Excessively warm, humid conditions, on the other hand, are undesirable. Insect pests do not seem to offer any problems to successful cultivation, but fungous organisms may present problems from time to time, particularly soil-borne ones. When planting *P. nana* it will pay to select soil that has not been heavily cropped or had other plants die in the situation.

PROPAGATION:
Can be reproduced from either seed or semi-hardwood cuttings. The latter is more difficult, however, and for the amateur, seed-raising will prove less frustrating. Seed normally germinates within a month from sowing.

USES:
Probably one of the most sought-after proteas by enthusiasts, *P. nana* has developed an aura of rarity and exclusivity about it. Due in no small part to its rather dainty beauty, being loved by gardeners and enthusiasts alike, but primarily due to its scarcity in garden centres and nurseries, this aura is likely to continue for some time.

As a talking-point in a garden, as a delightful subject for posy-bowls and similar small floral works, and to provide a soft, lighter green foliage effect in a shrub border, *P. nana* is certainly desirable. As a protea for commercial picking purposes, *P. nana*, is unlikely to have any future, the pendulous flowers not being sought after by many florists. In spite of this, it is hoped that this charming species will become more freely available to gardeners as time goes by.

PLATE 102: *Protea nana*.

Zoë C

PROTEA NERIIFOLIA

Commonly known in South Africa as the Oleander-leaved Protea due to the foliage similarities with that shrub. This is a very widely distributed species and is commonplace throughout South Africa's southern Cape. It is mainly found in acid, sandy soils at altitudes ranging from sea level to 1300 m. The climate is generally mild with a regular rainfall. Light frosts are occasionally experienced. *Neriifolia* means nerium (oleander)-like foliage.

DESCRIPTION:
Growth habit: in most instances an erectly bushy shrub to 2 or 3 m. Particularly old specimens may reach a height of 3 m with a spread of nearly as much. Some variants may assume a lower, spreading growth habit, but this is the exception rather than the rule.
Leaves: generally a fairly bright, medium green colouring, sometimes with noticeable red margins. Always narrow and elongated, measuring anything from 100 to 180 mm in length but only 15 to 30 mm in width.
Flowers: very variable, ranging from pale, creamy greens through soft pink to quite deep reds, but always with contrasting bearded tips to the bracts. The typical *P. neriifolia* grown in gardens is a medium, clear pink colouring with clearly defined black tips to the bracts. The central flower mass is usually a clear cream or biscuit colouring. The main flowering period of the widely grown form is from autumn through winter into early spring. Occasional flowers may also be seen at other times of the year. Stem length is good at 300 to 750 mm.
Variants: multitudes of variants occur with *P. neriifolia.* Not only does it produce all manner of variations when grown from seed, but a wide range of hybrids are also known.

CULTIVATION:
Along with *P. repens* and *P. obtusifolia, P. neriifolia* is one of the easiest and least demanding proteas to grow in the average home garden. Tolerance of alkaline soils, light sandy soils, heavier clays, a fair degree of coastal wind, and some frost, plus providing an abundance of flowers over an extended period, all combine to make *P. neriifolia* a firm favourite with gardeners. Ideally, a well-drained soil that is basically acid in character, plus a sunny position that is not subjected to excessive frost are the main requirements for successful cultivation of this popular species. Other than the infrequent attentions of caterpillars, which may cause some foliage damage, very little problem is experienced with pests and diseases. Finished flowerheads should be cut off, either with a short length of stem or without, to encourage plentiful new growth, which will tend to keep the bush compact in habit, as well as longer flower stems.

PROPAGATION:
P. neriifolia may be propagated by either seed or cuttings. Seed, as already mentioned, will give considerable variation. Ideally, forms exhibiting desirable characteristics of flower colour and growth habit should be propagated by taking cuttings of semi-hardwood in either late spring or autumn.

USES:
One of the showiest proteas for display over a long period, and being one of the easiest to grow is deservedly popular. Blooms of selected forms are frequently used both for commercial picking and for floral art purposes by enthusiasts. In landscaping, *P. neriifolia* is usually used as a taller specimen shrub or in mixed border plantings.

PLATE 103: *Protea neriifolia,* typical variants.

PROTEA NITIDA

Widely distributed throughout the south-western Cape, this species may be found growing from rocky shorelines where it is subjected to strong, salt-laden winds right through to higher, mountainous areas where snow is common in winter. Soils are usually either rocky or sandy and essentially acid in character. Well-drained, sunny slopes are generally preferred. It was previously known as *P. arborea*. *Nitida* means shining and *arborea*, tree. Commonly known as waboom.

DESCRIPTION:
Growth habit: *P. nitida* grown from seed is generally a bushy, tree-like shrub of 3 to 4 m in height. In the natural habitat, dwarf, multi-stemmed forms have been noted, while old specimens have been recorded at 6 or even 7 m high.
Leaves: a feature of this species, they are large, leathery, glaucous (blue-green) in colour and may be up to 180 mm long and 60 or 70 mm wide with a reddish margin. Young growth may be a bright red.
Flowers: unusual bracts are a soft yellowish-green and surround the prominent central flower mass of silvery-white or, rarely, pinkish-red colouring. Neat at the bud stage and upon first opening, it quickly becomes untidy as the central flower mass collapses. Usual flowering time is from autumn through winter and into early spring. Occasionally flowers can be seen at other times.
Variants: as mentioned, most examples in cultivation are typical of the species, being fundamentally small trees in character and all having the same flower type. In nature, a form with a prominent red centre has been observed, but it is rare. Dwarf, multi-branched forms are also recorded. Some forms exhibit young growth that is particularly bright red in colouring.

CULTIVATION:
A relatively easy protea to grow in gardens, provided the soil is well drained. *P. nitida* appears to be tolerant to the odd heavier frost, and certainly is resistant to salt-winds, a feature that should endear it to coastal gardeners who experience salt-burn on other species. As a rule, this species does not seem to be subject to any particular pests and diseases.

PROPAGATION:
While some success can be achieved with semi-hardwood cuttings, it does not appear as easy as some species to propagate by this method. Some viable seed may be found in established plants, and generally any plants offered for sale by nurserymen will probably have been grown from these.

USES:
An attractive larger shrub or small tree for bold landscape uses, the distinctive foliage showing up well and contrasting strikingly with other shrubs. Some floral artists have found the flowers and buds useful for floral work, but as a rule it will probably find little favour as a commercial picking species due to the fairly quick collapse of the flowers, becoming untidy within a few days of picking. *P. nitida* certainly has a place in any protea collection, however.

PLATE 104: *Protea nitida.*

PROTEA OBTUSIFOLIA

The most southerly-growing protea, being found in the extreme southern coastal regions of the Cape. It is one of the only proteas to grow almost exclusively in naturally alkaline soils that are either limestone or derived from limestone. Strong salt-laden winds are commonplace for much of the year in the exposed habitat. Originally known as *P. calocephala*, which translates to 'beautiful head'. *Obtusifolia* means round-tipped leaf.

DESCRIPTION:
Growth habit: forms a large, rounded shrub or small tree of up to 3 m with maturity. Old specimens of 20 years or more may be seen in gardens, and in Australia one particularly old bush is reputed to be 75 years old.
Leaves: particularly distinctive, being dark green, broad and quite glossy, although slightly hairy at the younger stages. Size is generally in the region of 100 to 150 mm long and 20 to 40 mm wide.
Flowers: variable in colour as illustrated, ranging from pinkish shades to brilliant lipstick-red colourings with creamy-white at the opposite end of the spectrum. The brighter red colour forms are the usual, although the other shades are occasionally seen. Size is up to 120 mm long and to 80 mm in diameter when open. Flowering usually commences in autumn and continues until spring or early summer. Mid winter is the peak period. Stem length is good, usually being in the vicinity of 300 to 500 mm.
Variants: as already mentioned, flower colour is variable, but little if any variation occurs in growth habit. Under cultivation in South Africa, some attractive hybrids have been observed, in particular with *P. compacta*.

CULTIVATION:
Equally at home in alkaline or acid soils, tolerant of heavier clay soil types, able to tolerate salt-laden and gale force winds better than almost any other protea, *P. obtusifolia* is certainly an adaptable species and suitable for trying in difficult areas where other proteas may fail. Frost can cause problems, heavier ones proving fatal to this species which has a relatively frost-free natural environment. Pests and diseases do not usually present any problems to the successful cultivation of this species. The odd caterpillar may cause some leaf damage, but this is not normally a problem of any significance.

PROPAGATION:
The majority of plants sold by nurserymen have been seedlings, but cutting-grown, selected forms will probably be the main ones offered in the future. Cuttings should be taken from semi-hardwood in late autumn or spring as a rule.

USES:
A handsome protea for a variety of uses, *P. obtusifolia* is ideally suited to coastal gardens where wind burn is a problem. In strongly alkaline soils where other proteas fail, it is bound to succeed — although it is equally at home in average, acid soil types. Colourful flowers over a long period during winter make it popular with many gardeners. As a cut flower it is somewhat underrated, as it does have a good vase life and is a colourful flower for a variety of arrangement styles.

PLATE 105: *Protea obtusifolia*, typical variants.

PROTEA PENDULA

Found growing in a restricted mountainous area of the central south-western Cape in South Africa, at altitudes of between 1000 and 2000 m. The climate tends to be hot and arid over summer with cold winter temperatures, snow being commonplace. Rainfall is low at 300 mm or less. Soils are rocky, well drained and acidic. *Pendula* means hanging down.

DESCRIPTION:
Growth habit: ultimately forming a dense, thickly branched shrub, erect in habit to a maximum of 2.5 m in nature, with a spread of almost equal dimensions. Regarded as slow growing in its natural environment, only reaching its prime after 15 to 20 years and living to 50 years or more. Eight-year-old specimens in the author's garden have reached heights of 1.8 m and have a spread of 1.5 m. They flowered after 4 years.
Leaves: distinctively grey-green in colour, covered with a fine, suede-like pubescence at first, this becoming less noticeable with maturity. On the flowering stems, the foliage tends to reflex backwards, away from the bloom. Dimensions of the mature leaves are usually about 10 to 15 mm wide and 30 to 50 mm in length.
Flowers: always pendulous (hanging down) on gracefully arched stems. The main flowering period is over winter, with odd flowers occasionally opening up until late spring. The backs of the usually red bracts are attractively dusted with a fine silvery pubescence. Bract colour is usually an overall red shading and the central flower mass is mainly a tawny brown. Size of the shallow bowl-shaped bloom is 40 to 70 mm across and only 30 mm in depth. Stem length is in the order of 100 to 150 mm.
Variants: more common are the deep red shadings but variants are seen with bract colouring ranging from green through pinks to dark reds.

CULTIVATION:
Regarded as a difficult species to grow successfully except in cold and dry inland areas. The author's 8-year-old plants have had no special attention, have flowered regularly for the past 4 years and apart from a little fungous blackening of the foliage have not exhibited any special problems. They are growing happily on a slight slope, in light, acid soil and receive maximum sun. In line with their natural habitat, they have not been affected by frost. However, in total contrast to the natural environment, they are in a fairly mild, coastal region with rainfall of over 1000 mm on average.

PROPAGATION:
As plants sold seem true to type and have been propagated from seed, it would appear that this is a reliable method. Cuttings may be taken from semi-hardwood in late spring or autumn, but are more difficult to strike than most other species.

USES:
Although a very lovely protea, because the blooms hang their heads it will probably never be a popular species with the home gardener. As a collector's plant it is well worth growing. The bushes are neat and have attractive foliage, and the flowers are most appealing. Their potential for picking purposes is limited, but for smaller posy-bowl work they could have some use.

PLATE 106: *Protea pendula.*

Zoë C

PROTEA PITYPHYLLA

Appropriately known in South Africa by the common name Long-leaved Mountain Rose, this species is regarded as rare in nature. Found in only limited numbers, usually in small colonies of only a dozen or so plants, over a very restricted area of the south-western Cape. The habitat is usually in mountainous areas of 500 to 1500 m, among large sandstone boulders. Fairly cold winters are the norm. The Latin *pityphylla* translates to pine leaf.

DESCRIPTION:
Growth habit: low and spreading, generally rarely exceeding 600 or 700 mm in height, although old specimens in nature have reached 1 m high. Spread may be up to 2 m, although it is rare to see cultivated plants much more than 1 m across.
Leaves: very narrow and needle-like, similar to *P. nana* only longer and more prominent, particularly just behind the flower, where a marked concentration occurs. Medium to deep green, the leaves may measure 90 mm in length and only 1 mm wide.
Flowers: similar in shape and colouring to *P. nana* and pendulous, as with that species. They are, however, considerably larger, being up to 80 mm across the shallow, bowl-like bloom. The bract colour can be described as a deep rose or mahogany-red, while the central flower mass is pale gold, tipped with bronze-red. Main flowering time is spring, but blooms may be seen from early winter onwards. The long flowering stems which bear the pendent blooms are almost horizontal with the ground and can be up to 200 or 300 mm in length.
Variations: as most plants sold are grown from seed, some slight variation in flower colour may occur, but most are true to type. In nature, hybrids between *P. pityphylla* and *P. effusa* have been noted.

CULTIVATION:
Regarded in South Africa as a robust, vigorous grower. Because of its low, spreading habit, it is a desirable rockery plant and would also be suitable for growing on a bank where it can be seen to best advantage. Well-drained soils that are acidic and basically light in character should be suitable. Light to medium frosts should cause no problems. No particular pest or disease problems have been observed with this species.

PROPAGATION:
As mentioned, this species is usually grown from seed which needs to be fairly fresh to retain its viability. Little information is available on vegetative propagation, but it is more difficult to grow from cuttings than other proteas.

USES:
As a rockery or bank subject, *P. pityphylla* is a very ornamental species. It may be picked for floral work, but has probably little potential commercially.

PLATE 107: *Protea pityphylla*.

PROTEA PUDENS

Commonly known as Ground Rose in its native South Africa, and still widely known by its former botanical name of *P. minor*. Occurs naturally in a very limited area of the southern Cape, mainly in soil of a heavy, clay type, often overblown with sand. The climate is generally mild with a low winter rainfall. *Pudens* means bashful.

DESCRIPTION:
Growth habit: low and spreading, attaining a height of little more than 400 mm, *P. pudens* may reach a total spread of a metre or more.
Leaves: narrow, grey-green, usually from 60 to 100 mm long and only 2 to 5 mm wide.
Flowers: similar in appearance to a miniature form of *P. longifolia* to which it is closely allied. Blooms are nearly always a dusky, deep rose-red shade, with a central peak to the flower mass which is variable from black to red and purple shadings. Size is quite small, being 50 to 80 mm long and 40 to 60 mm wide. Stem length is generally short at 120 to 200 mm. The main flowering period is over winter, with the odd flower sometimes appearing at other times of the year.
Variants: although no particular variants are recorded in South African literature, plants grown in cultivation appear to be quite variable in flower colour, ranging from very soft pinks to deep reds. This species is known to hybridise in nature with *P. longifolia*, and plants have been marketed as *P. minor* hybrids during the late 1970s and early 1980s. While these have usually exhibited some of the characteristics of *P. pudens*, evidence of *P. neriifolia* parentage can also be observed.

CULTIVATION:
Generally regarded as a very adaptable and easily grown protea, *P. pudens* is tolerant of alkaline soils and light frosts and will grow well in most average garden situations. Few if any pests and diseases seem to affect this species under cultivation. The occasional caterpillar hole may blemish foliage, but this is not a problem.

PROPAGATION:
Plants obtained from seed may be variable, hybrids often occurring. Plants exhibiting good colour and growth habits should always be propagated by taking cuttings from semi-hardwood, usually in late spring or autumn.

USES:
Suitable for the foreground of shrub borders, where its lower habit is not lost amongst taller shrubs. A delightful collector's item. Flowers may be picked, and while it probably has no particular commercial value, floral artists find it a useful and very attractive smaller protea.

PLATE 108: *Protea pudens.*

PROTEA PUNCTATA

A widely distributed protea, being found throughout the drier, inland mountain ranges of the southern Cape at altitudes of 1200–2000 m. Situations are normally dry and sunny with well-drained, acid soils. *Punctata* translates to marked with dots.

DESCRIPTION:
Growth habit: forming a tall, rounded shrub of up to 3 or 4 m in height, under cultivation old specimens may sometimes be seen as rather sparsely branched and gnarled looking shrubs of less than 2 m.
Leaves: broadly oval in shape, deep blue-green in colouring, younger foliage having a silken pubescence. Size of the leaves is variable, often considerably larger in moister climates, but usually in the region of 40 to 80 mm long and 20 to 40 mm wide. The general appearance of foliage is thick and leathery.
Flowers: commencing with a distinctive, cylindrical bud of 70 or 80 mm long and 20 mm wide, the bracts open out at right angles to the central flower mass. At this stage, the whole bloom resembles a water lily. Gradually the central flower parts collapse and the fully open flower soon becomes untidy in appearance. At the open stage the overall width may be 80 or 90 mm. Stem length is usually 200 to 250 mm or in well-grown younger plants up to 300 mm. The main flowering period is from mid summer to the beginning of winter, with a peak in the autumn months. The usual colour of the better forms is a rich shell pink, but both deeper and lighter versions are sometimes seen.
Variants: Many variations of *P. punctata* are offered by nurserymen. Some are actually *P. lacticolor* or *P. mundii*, while others are thought to be hybrids. An ivory coloured form is sometimes offered by nurserymen.

CULTIVATION:
In South Africa, *P. punctata* is regarded as a temperamental species best suited to dry, well-drained sites in full sun, and not very easy to cultivate in the humid, higher rainfall areas. In other countries, however, as with so many of the proteas, *P. punctata* appears quite adaptable to most regions. It seems tolerant of the moister and more humid coastal climates as well as the colder and drier inland areas. As long as the soil is well drained and the situation sunny, many gardeners seem to succeed with this species, and it does not appear to have any particular pest or disease problems.

PROPAGATION:
P. punctata is easily raised from seed, which as a rule germinates quite quickly. Some variations are likely to occur with this method of propagation, however, and if a desirable form is required then cuttings should be taken from a selected stock plant. Floral artists are most particular as to the 'correct form', and in their eyes the form illustrated here is the right one.

USES:
As already indicated, *P. punctata* is particularly popular with floral artists, the shell-pink colouring and delicate form lending itself well to a variety of arrangements. Its use as a commercial cut flower is probably limited by its fairly short vase life. This can be extended by picking at the optimum bud stage, i.e., when the bud is soft and the brown of the tips of the central flower mass are just showing above the pink bracts.

As a garden plant, this species retains a fairly neat, erect shape, or at least can be encouraged to do so by a little judicious pruning. The delightful flowers are carried over an extended period and it creates an attractive display in the shrub border.

PLATE 109: *Protea punctata*.

Zoë C

PROTEA RECONDITA

This unusual and striking species is found only in mountainous areas of the south-western Cape at altitudes of 1000–2000 m. Winters are cold with some snow commonplace, with summers hot and dry. East-facing slopes of high peaks are the preferred sites. In such bleak situations plants are often found rooted into small crevices in rock faces from which they sprawl. In other situations plants are found in dense groups in open veld. The Latin *recondita* means hidden, referring to the flower being encased by the foliage.

DESCRIPTION:
Growth habit: Low growing and spreading, old bushes in nature reaching a maximum height of 1 m but bushing out to 2 m or more. Three-year-old plants under cultivation have reached 1 m across but are still only 300 mm high.
Leaves: handsome, glaucous, blue-green foliage, often with pinkish tints during the winter months. Quite prominently veined and up to 160 mm long and 55 mm wide, they are a feature of the plant when it is not in flower.
Flowers: normally a pale greenish-yellow colouring with a faint red margin to the bracts, the one illustrated is a deeper shade with more pronounced red margins. Quite large, to 90 mm or more across, of broad cup-shaped appearance and sometimes virtually encased by the foliage. Flowering commences with the onset of winter.

CULTIVATION:
Regarded in South Africa as being an unsuitable species for cultivation in mild, coastal districts, and more suited to the colder inland areas. Most well-drained soils that are not excessively high in nutrients should be suitable for successful cultivation. In warmer districts, some problems with leaf fungous can occur, and this is controlled with the use of a copper or systemic fungicide.

Obviously this protea will prove perfectly happy in districts where heavy frosts occur.

PROPAGATION:
From seed or cuttings taken from semi-hardwood in late spring or autumn.

USES:
A rather intriguing and different protea for foreground planting in shrub borders, for spilling down banks or simply to add to enthusiasts' collections. The beautiful foliage and the large cup-shaped blooms are almost certain to cause *P. recondita* to become a sought-after and prized species.

PLATE 110: *Protea recondita.*

PROTEA REPENS

This species has been widely cultivated as an outdoor garden plant in many countries outside its native South Africa since the beginning of this century — particularly in New Zealand, Australia and the United States (in California). Commonly known as Sugar Bush in its native land, due to the copious quantities of nectar exuded by the flowers. This is the most widely distributed of all Cape proteas and the most common South African species. It may be found growing near the sea in sandy soils to lower mountain slopes in rocky ground. Soil conditions may also vary from very acid to alkaline and may be either very light and well drained or poorly drained, heavy clay. In some localities, quite severe winter frosts are experienced, while in others almost subtropical, humid conditions are the rule. *Repens* means creeping, but in this case the name does not reflect the habit. Formerly *P. mellifera*.

DESCRIPTION:
Growth habit: generally an erect-growing, bushy shrub to 2 m or more, although some forms may tend to sprawl outwards before becoming erect growing, giving a broad, dense shrub of greater width than height.
Leaves: variable, but generally long and narrow, ranging from 50 to 120 mm long and from only 5 to 18 mm wide. Colouring ranges from a bright shiny green to a duller blue-green shade.
Flowers: buds are usually 100 to 120 mm long before opening to form a wide V-shape like a shuttlecock. Fully open flowers may measure up to 150 mm across, but more usually 90 to 100 mm. They quickly become sticky with the abundance of nectar that is exuded, a feature of the flower that attracts bees and nectar-seeking birds alike. Flowering time is variable, according to the area from which the plants have originated. A selection of *P. repens* forms can provide flowers throughout the year.
Variants: as already mentioned, considerable variation occurs among plants of *P. repens*. This may take the form of differences in growth habit and flowering time, but also in flower colouring. This may vary from almost pure greeny white to delicately shaded white tipped with soft pink to quite gaudy scarlet and cream-toned forms, and finally almost pure, rich red.

CULTIVATION:
Undoubtedly one of the easiest proteas to grow, *P. repens* is noted for being tolerant of widely varying soil types, from sand to clay, from acid to alkaline, from poorly drained types to dry, very free-draining soils. It is also quite resistant to salt winds which should make it very suitable for coastal gardens. Hardiness to cold is dependent on the region from which plants have originated, and it is unfortunate that plants sold are often not identified in this way. Few insect pests seem to trouble *P. repens*, but it is subject to some foliar fungous diseases which can cause blackening of young growth. This can be controlled with the use of systemic or copper fungicides.

PROPAGATION:
Simple to propagate from seed or cuttings, the former sometimes giving rise to interesting hybrid forms (see *P.* 'Clark's Red' page 190).

USES:
A reliable, long-lived protea for many garden situations. Gives bloom over an extended period, often through the bleak early winter months. Several forms of *P. repens*, particularly the cream-flowered types, are popular for floral work.

PLATE 111: *Protea repens*, colour variants.

PROTEA ROUPELLIAE

Found in the eastern half of South Africa, *P. roupelliae* occurs in the summer rainfall areas and is therefore distinct from most of the other commonly cultivated Cape proteas which occur in winter rainfall regions. It grows in widely divergent soil types, from limestone to sandstone, and either alkaline or acidic. It is to be found in areas of both subtropical heat and extreme winter cold. *Roupelliae* — after A. E. Roupell (1817-1914), an artist and author of *Specimens of the Flora of South Africa, by a Lady* (1849).

DESCRIPTION:

Growth habit: forms a small tree of anything from 3 to 8 m in height and may assume a gnarled appearance in old age. Generally a long-lived species.

Leaves: lush, soft, bright green in colour, 100 to 150 mm in length, 25 to 40 mm wide, slightly hairy on the margins. Some variants may be more grey-green and more pubescent.

Flowers: bracts are open and widely spaced, showing the inner flower between them. Colour is a soft pink as a rule, deepening towards the bract tips. The apex of the central flower mass is shaded into the violet tonings. Usual flowering time is over late summer. Stem length may be up to 400 mm.

Variants: in nature a semi-prostrate subspecies, *P. roupelliae* ssp. *hamiltonii* is recorded, but is rare. Plants seen in cultivation vary in foliage and flower colouring but are still usually recognisable as the species. Our illustration depicts an exceptionally lovely form which is possibly a hybrid of *P. roupelliae* and *P. eximea,* a cross which is not uncommon in South Africa when grown under cultivation.

CULTIVATION:

A reliable and easy protea to grow in many gardens. Most soils would appear to be suitable as it is remarkably tolerant of varying types. Very few pests or diseases appear to affect this species.

PROPAGATION:

May be propagated from seed, which can give rise to some variations. Cuttings taken from semi-hardwood in late spring or autumn are reasonably easy to strike, and ensure that plants will have the same characteristics as the parent stock bush.

USES:

A handsome small tree for landscape uses, creating natural background for other protea plantings. The foliage is attractive when the bush is not in flower. Flowers may have some limited use for picking.

PLATE 112: *Protea roupelliae,* a variant that may be a hybrid with *Protea eximea.*

Zoë C

PROTEA ACUMINATA

Formerly known as *P. cedromontana* and commonly known in its native South Africa as the Cedarberg Protea. This species occurs in three separate and distinct areas of the south-western Cape — ranging from one which is arid with snow in winter to one with three times as much rain and an equable climate. Soils are generally sandy, well drained and acid. The more robust forms are found in the most arid region. *Acuminata* means pointed leaves.

DESCRIPTION:
Growth habit: usually a neat, small bush with an erect habit to 1 or 1.5 m. In nature, very old specimens have been observed that have a height and spread of 2 m.
Leaves: long and narrow, 60 to 120 mm long and only 5 to 8 mm wide. Medium to deep green, sometimes with a greyish tinge. Frequently conspicuously surrounding and initially enclosing the flower buds.
Flowers: may be either deep mahogany red or lighter red and green. Always distinctively cup-shaped and varying in diameter from 35 to 60 mm. Late winter is the peak flowering period.

CULTIVATION:
Colder, drier inland areas are generally preferred, moist and humid climates often leading to fungous problems. It is usually frost tolerant.

PROPAGATION:
Seed usually produces plants true to type. Cuttings produce plants that flower earlier and of course are always true to the parents' characteristics.

USES:
An ornamental small species suitable for the garden in cooler climates. Nice picking for posy-bowl work.

PLATE 113: *Protea scolymocephala,* (upper right), *Protea acuminata,* (lower left).

PROTEA SCOLYMOCEPHALA

Commonly known in South Africa as Small Green Protea, and by many gardeners and florists in other countries as 'Scoly'. Occurs naturally at lower altitudes (usually less than 100 m) in the coastal regions of the south-western Cape. Soils are mainly of a deep sand type, but in a few locations this species is found growing on heavy clays. The climate is generally mild. The Latin *scolymocephala* refers to the thistle-like (scolymus) head.

DESCRIPTION:
Growth habit: a dense, rounded shrub with a short trunk, reaching a maximum height of 1.5 m with an equal spread. Specimens of 8 or 10 years old become more sparsely branched and eventually become rather open and gnarled in appearance.
Leaves: carried densely on the short branches, medium green in colour, small and narrow, 25 to 50 mm long as a rule and only 3 to 6 mm wide. They have a noticeably sharp point to the apex, which is usually coloured red or black. Often more thickly clustered at the base of the flowers.
Flowers: carried in profusion over late winter to summer, the peak being reached over the spring months. Mature bushes may produce several hundred of the small, bowl-shaped blooms. Overall colouring is a creamy green, the backs of the flower bracts sometimes having distinctive red markings. The tips of the central flower mass are a silvery grey shading. Size of the open flower is up to 45 mm in overall width. Stem length is rarely more than 250 mm.
Variants: with plants grown from seed, some variations in flower size and form have been observed but as a rule with most plants being offered as cutting grown specimens, they are a uniform type.

CULTIVATION:
One of the easier proteas to grow, *P. scolymocephala* appears to adapt well to most gardening regions. Once established, it withstands both drought and cold well (to −6 °C). Although it appears to grow quite successfully on heavier soils, it is undoubtedly happiest in lighter sandy soils. Coastal gardens seem well suited, and as a rule this species stands up well to windy

conditions. As with most proteas, full sun is required, and soils that are basically acid are preferred. No particular pests or diseases seem to favour *P. scolymocephala*.

PROPAGATION:
While this protea can be quite easily propagated from seed, most plants sold are cutting grown. The latter method seems to give strong plants which are quicker to flower, and as cuttings are easy to strike it seems to be the preferred method. Material should be taken from healthy bushes, selecting semi-hardwood which is normally at the right stage in late spring or early autumn.

USES:
A very popular protea, especially with floral artists and florists who find the little blooms ideally suited to posy-bowl work. A reliable shrub for the home garden, it quickly forms a dense bushy specimen that fits into shrub borders nicely. Although the blooms are not spectacular in themselves, the sheer number produced makes the bush quite eye-catching. Certainly a very worthwhile protea for a variety of purposes.

PROTEA STOKOEI

Found only in a restricted area of the south-western Cape, in fairly mountainous terrain, this species occurs in areas of mild climate, high rainfall and humidity and in very acid, well-drained soils. Named for Thomas Stokoe (1868–1959), an amateur botanist responsible for discovering many new South African species.

DESCRIPTION:
Growth habit: growing to 2 m in height, erectly bushy in habit, this species is always neat in appearance. Old bushes may reach to 3 m in ideal conditions.
Leaves: broadly oval in shape, thick and leathery with a conspicuous red margin, basically a deep green colouring. May be 80 to 120 mm long and 50 to 70 mm wide.
Flowers: unusual and distinctive. Soft, brownish pink in colour, each bract being tipped with a dense mass of brown fur. The blooms never open completely, showing only a small tuft of cream inner flower. Main flowering time is winter and spring. Stem length is rarely in excess of 30 to 40 cm.

CULTIVATION:
A species requiring a soil that is acid and free draining but with a relatively high humus content. The addition of organic material such as peat or decayed compost should assist with successful cultivation. It is unlikely that *P. stokoei* will tolerate heavy frosts as it naturally comes from a mild climate area. No particular pest or disease problems have manifested themselves to any degree under garden conditions.

PROPAGATION:
Most plants sold are grown from seed, but some success can be achieved with vegetative propagation, using semi-hardwood taken in late spring or autumn, depending on the season and the type of wood produced.

USES:
A delightful and unusual protea that should be grown more. It is neat and of modest dimensions, making it suitable for smaller gardens. It should respond well to garden conditions, needing a reasonable amount of moisture, organic material, and acid conditions. The flowers are particularly attractive and are useful for picking purposes, having a good vase life. It is interesting to note that in South Africa *P. stokoei* is both slow growing and slow to flower, and considered generally difficult to grow. In some other countries it flowers more quickly, is faster growing and gives few problems.

PLATE 114: *Protea stokoei.*

PROTEA SULPHUREA

Found growing in several inland areas of the south-western Cape. The habitat is almost exclusively arid, receiving virtually no rain over summer. Snow is experienced during winter, and the terrain is mountainous at altitudes of 1000–1550 m. This species forms broad, mounded hummocks on flatter parts of the veld, or by contrast cascades from improbable crevices on cliff faces in the steeper country. Soils are usually poor and well drained. Commonly known in South Africa as Sulphur-coloured Protea. Although first discovered in 1903, it was not generally cultivated until the 1950s, either in South Africa or elsewhere. *Sulphurea* means sulphur coloured.

DESCRIPTION:
Growth habit: low and sprawling or 'mounding' in appearance. May reach a height of 500 mm, with a spread of 2 m or more.
Leaves: small for a protea, usually glaucous green in colouring. Approximately 30 mm long and 10 mm wide, with tiny pointed tips of either black or red.
Flowers: may take up to 7 years from seedlings and 3 years from cutting-grown plants for flowers to appear. Buds appear during summer and open to form spectacular flowers through winter. Shaped like pendulous, shallow bowls up to 130 mm wide, the blooms are carried like glowing lanterns. The reverse of the overlapping bracts is a bronze-green shade, edged with red. Inner surfaces of the bracts and the base of the central flower mass is a strikingly beautiful sulphur yellow. The tops of the central flower are a deep golden-bronze colour. Open flowers are reputed to emit a strong scent, variously described as being either 'yeast' or 'honey-like'. The blooms produced on the author's plants do not have this characteristic.
Variants: in nature, variations occur with locality. Some plants may have almost pure red backs to the flower bracts. Others have larger or more green foliage.

CULTIVATION:
An easy species to grow in lighter, well-drained soils. An open, sunny position is desirable. As the flowering habit is peripheral, i.e., blooms are formed towards the outer growth of the plant, the ideal situations for successful flowering are banks and walls where *P. sulphurea* can hang or cascade. Particularly on younger plants, formation of buds is 'aborted' where outer growth comes into contact with the soil. With maturity, when bushes mound, this problem is less likely to occur. The author's plants, which flowered profusely at 3 years old, are hanging over retaining walls.

Observations lead to the conclusion that this species thrives in soils that are naturally very low in nutrients. Therefore all fertilisers should be avoided.

Insect pests do not appear to be a problem, but occasional tip dieback caused by a fungous disease may be experienced in humid weather conditions. This could be countered with the use of a copper or systemic fungicide.

PROPAGATION:
By either seed or semi-hardwood cuttings taken in late spring or autumn. Seedlings are normally true to type, but as mentioned may take considerably longer to flower.

USES:
An excellent species for landscaping situations where bank cover or a feature to hang over a wall is required. The colour of the foliage and general appearance of the bush with its dense, neat habit is most appealing. For the *Protea* enthusiast or connoisseur, *P. sulphurea* must be considered an essential addition to a collection. The flowers are outstandingly beautiful and the clear, sulphur-yellow colour is not often found in proteas. From a picking point of view, the pendulous habit may be considered a disadvantage; florists no doubt would consider wiring necessary to artificially straighten the stem. Floral artists on the other hand may well consider the flowers of considerable merit for their arrangements.

PLATE 115: *Protea sulphurea.*

PROTEA SUSANNAE

A coastal species mainly found in close proximity to the ocean, often growing in deep sand but also on limestone, it is not usually found at altitudes of more than 100 m. It is found in a narrow coastal strip of the south-western Cape. *P. susannae* is one of the few proteas to be found growing on alkaline soils as well as acid ones. Named for Susanna Muir (1882–1970).

DESCRIPTION:
Growth habit: a dense, bushy shrub growing to 3 m in both height and width.
Leaves: of typical protea shape, being basically lanceolate, dull green, 100 to 150 mm long and 20 to 30 mm wide, quite prominently veined and sometimes with a faint red margin. When damaged, the leaves have a distinctively unpleasant odour.
Flowers: a neat, reddish-brown bud forms from late summer and during autumn, and through winter and into spring opens to form a compact, pinkish-red flower with brown tints. The central flower mass is a pinkish-white colour and becomes untidy when the flower opens up.
Variants: most *P. susannae* plants sold are cutting grown and therefore exhibit the same characteristics. Prior to this, when plants were grown from seed, some variations were seen to occur, these mainly being in the intensity of flower colour, but some plants that were obviously natural hybrids of *P. susannae* and *P. eximea* have been observed.

CULTIVATION:
Probably the easiest protea to grow. It withstands a wide range of soil types, from very acid to markedly alkaline. Salt-laden winds in coastal gardens do not present a problem to successful cultivation. Any average soils that are well drained should be satisfactory, sandy soils being particularly suitable. Heavy frosts in inland situations may cause losses, and in such conditions protection will be required.

PROPAGATION:
Usually propagated from cuttings of semi-hardwood taken in late spring or autumn. May also be raised from seed, but some variation may occur, with the resultant plants not always exhibiting the full characteristics of the parent bush.

USES:
A robust and reliable protea for landscaping uses where a larger specimen is required. Ideal for sandy banks to provide broad, wind-resistant cover. Flowers for a long period over winter. Not particularly suitable for picking purposes unless the arrangement is such that the leaves may be stripped to avoid the unpleasant smell. A very suitable protea to try where others may have failed, as it is tolerant of such a broad range of soil types.

PLATE 116: *Protea susannae.*

PROTEA TENAX

Found growing in a restricted area of the southern Cape in South Africa, but in regions varying remarkably climatically. Some habitats are very arid, receiving low rainfall, while others are quite moist, receiving good rain regularly. The altitude range in the mountainous country where *P. tenax* occurs is from 1250–1750 m, accordingly winters are cold. Soils are generally poor, acid and well drained. Formerly known as *P. canaliculata*, which confusingly is now the name for what was *P. harmeri*. *Tenax* means tough.

DESCRIPTION:
Growth habit: prostrate or sprawling in habit, reaching little more than 100 mm in height, but spreading to a metre or more. It has long, trailing stems on which the flowers are borne.
Leaves: variable, but most forms are a medium, glossy green colour, long and narrow and generally measuring 100 to 140 mm in length and only 10 to 25 mm in width.
Flowers: rounded and shallowly bowl-shaped, 60 to 70 mm in diameter. Delicately coloured in creamy-green with hints of red. Borne on the end of the long, trailing stems, mainly during winter and early spring.
Variants: in nature varying forms occur with different habitats. Flower colour varies slightly, but the main differences are in the leaves, which vary markedly in width, length and colour — some being almost a glaucous blue-green.

CULTIVATION:
P. tenax is undoubtedly a species suited to growing in a rockery or cascading down a bank or wall. Its sprawling, prostrate habit lends itself to these applications. Soil that is not excessively high in nutrients, on the acid side and well drained is most suitable. Pests and diseases do not generally trouble this species, and the only problem noted is the occasional caterpillar hole in foliage.

PROPAGATION:
Comparatively simple to propagate, seed or semi-hardwood cuttings being equally satisfactory. Cutting-grown plants will prove to be more robust at an earlier stage and will usually produce flowers more quickly.

USES:
A particularly suitable protea for the situations previously outlined — rockeries, banks, walls and similar positions. Flowers have not been used to any degree by florists or floral artists, but this is probably due to a lack of material. *P. tenax* should prove a delightful flower for such purposes.

PLATE 117: *Protea tenax*.

PROTEA VENUSTA

This species occurs naturally at higher altitudes (1700–2000 m) among the mountains of the southern Cape. In winter, plants may become covered by snow. *P. venusta* may be seen rooted into crevices between rocks and in other unlikely situations and often cascades down banks or spills over rocky ledges. Obviously, soils are poor and well drained. *Venusta* translates as lovely or beautiful.

DESCRIPTION:
Growth habit: low and spreading, in nature reaching a width of up to 3 m with a height of little more than 700 mm. Dense in habit, and producing long, trailing flowering stems.
Leaves: particularly neat and carried densely on the shrub. The colouring is a blue-green and the size 25 to 40 mm long and 10 to 20 mm wide. The leaves often display a prominent red margin, adding to their appeal.
Flowers: mainly produced over late summer. Attractive, especially in the bud stage, when the still tight bracts are a strikingly deep pink shade, *P. venusta* flowers open to form a tight 'shuttlecock' shape of cream and pink colouring. When open, the flower bracts are soft cream at the base with the deep pink showing at the rounded tips. The inner flower mass is a uniform cream colour tipped with silvery-white. The size is 70 mm long and when open may be up to 100 mm across. Stem length is good at up to 300 mm. Unlike so many of the lower, sprawling proteas which tend to have either pendulous or partially concealed blooms, *P. venusta* holds its flowers prominently upwards giving a clearly visible display. It is interesting to note that in South Africa this species is regarded as being very shy to flower at lower altitudes and in coastal situations. This is not the case in every country and it may flower within 3 or 4 years and regularly.
Variants: although not mentioned specifically in South African literature, there appear to be two distinct forms grown. One, the more usual, is illustrated, while the other has larger, paler coloured flowers that are generally less attractive. In its natural habitat, hybrids between *P. venusta* and *P. punctata* have been observed, and they are regarded as being desirable as they flower profusely and retain the attractive flower shape and colouring of *P. venusta*.

CULTIVATION:
A very adaptable protea under most conditions, its ability to withstand heavy frosts and even snow making it useful for colder, inland gardens. Well-drained soils and open, sunny positions are the main requirements. Few if any pests and diseases pose any problem with this species.

PROPAGATION:
Probably propagation from cuttings is the most satisfactory method of attaining plants that are true to type. This does not present any problems, material from semi-hardwood striking readily and giving plants that are generally more robust and quicker to flower than seedlings.

USES:
An ideal protea for bank cover, the lower, dense, spreading habit making it ideal for such positions. Florally, the blooms are produced at a time of year when few other proteas are available. Vase life is a little limited, a week at the most, but when first opened from the bud it is particularly attractive.

PLATE 118: *Protea venusta.*

Comprising more than 44 species, the *Serruria* are one of the most unusual and intriguing genera of the Proteaceae.

SERRURIA FLORIDA

Once considered extinct, and in fact 'lost' for some 90 years, the lovely *S. florida* is still rare in nature being found in only one isolated, mountainous area of the Cape. The region has cool air with misty cloud prevalent, even in summer, and a rainfall of over 1000 mm. Soils are poor, derived from granite, and sometimes clay-like, although sandy or rocky soils on steep slopes are preferred. Commonly known wherever it is grown as Blushing Bride, a reference to the habit last century of young men wearing the flower in their buttonholes when courting. *Florida* is from the Latin *floridus*, profusely flowering.

DESCRIPTION:
Growth habit: in nature, where it is left unpruned, growth tends to be very erect and even 'lanky', to approximately 1.5 m. Under cultivation, plants that have been well pruned or picked develop into a more compact, bushy shape to about 1 m.
Leaves: medium green, delicately divided into long, narrow, almost needle-shaped leaflets to approximately 50 mm long. Overall foliage appearance is 'feathery'.
Flowers: the outstanding feature, white papery bracts, delicately flushed pink, surround the soft pink central flower mass. Flowering time is from mid winter through spring and the blooms are ideally picked at the stage depicted in the illustration.

CULTIVATION:
Well-drained, lighter soils that are acidic and not too high in nutrient levels are best suited. Warm, sunny positions and no more than the occasional light frost are also requirements for successful cultivation. Pinching out the growing point of young plants encourages bushy growth.

PROPAGATION:
All serruria species are relatively easily propagated by either seed or cuttings, the latter giving flowering plants more quickly, however. Cutting material should be of semi-hardwood taken in autumn.

USES:
An outstanding cut flower, the blooms having an almost legendary vase life. They also dry well. Commercially it is in danger of being 'overdone' with an oversupply being the case in many markets now that it is being so widely grown.

PLATE 119: *Serruria florida* (centre, left),
Serruria pedunculata (top),
Serruria barbigera (lower, right).

SERRURIA PEDUNCULATA

Commonly known as Grey Serruria and found growing on lower mountain slopes of the south-western Cape where soils are rocky, poor and well drained. Main rainfall is over the winter months. Formerly known as *S. artemesiaefolia*. Usually a low-growing bushy shrub to less than 1 m. Foliage is soft, delicately divided with a grey-green colouring. The small flowers, produced over spring, are a smokey, wine and grey combination, last well when cut, and are ideal for smaller arrangements and posy work. It is tolerant of light frost and is usually easily grown in well-drained soils.

SERRURIA BARBIGERA

Commonly known as Silky Serruria in South Africa. Found in a similar type of mountainous terrain to *S. pedunculata,* with varying forms found at different altitudes. *Barbigera* means bearded. Generally a low-growing, small, bushy shrub of only 60 cm. It has light green, finely divided, feathery foliage and produces silky little flowers of silvery pink and red over spring. It is inclined to be temperamental under cultivation and is easily damaged by frost. Good drainage is essential.

THE PROTEACEAE OF SOUTH AMERICA

The full number of species of Proteaceae found in South and Central America is not as well documented as those of Australia and South Africa, and the following list is given as a guide only.

Embothrium; 1 species.
Euplassa; 25 species (tropical America).
Gevuina; 1 species.
Lomatia; 3 species.
Oreocallis; 2 species.
Orites; 1 species.
Panopsis; 20 species (tropical America).
Roupala; 50 species (Central and South America).

EMBOTHRIUM COCCINEUM

Commonly known as Chilean Fire Bush. A small genus of shrubs and trees native to the Andes Mountains. *Alloxylon*, an Australian genus closely related, was until recently classified as *Oreocallis*, and prior to that as *Embothrium*. *Embothrium* translates to 'in a little pit', while the specific name comes from *coccineus*, meaning scarlet.

DESCRIPTION:
Growth habit: erect, slender tree to 5 m under cultivation, often to 12 m in nature. Tends to sucker and form clumps or thickets of dense growth if uncontrolled in the garden situation. In full bloom the weight of the flowers, particularly in the case of the cultivar 'Longifolium', may cause the branches to assume a weeping habit, often drooping to within a metre of the ground.
Leaves: lanceolate, deep green, 120 to 140 mm long and 20 to 30 mm wide. Narrower in the cultivar 'Lanceolatum'.
Flowers: dense clusters of honeysuckle-like blooms, orange-red in colour, carried in such profusion that the name 'Fire Bush' is very apt. The main flowering season is in spring and early summer, although odd flowers can be seen throughout summer and even into autumn in some districts.
Variants: the cultivars 'Longifolium' and 'Lanceolatum' are considered by some botanists to be geographical forms. *E. coccineum* 'Longifolium' is regarded as a more desirable garden subject due to its graceful habit, vigorous growth and better drought tolerance. *E. coccineum* 'Lanceolatum' is considered the more frost hardy, but is inclined to become almost deciduous, particularly in colder climates.

CULTIVATION:
Soils that are free draining but still moisture-retentive through adequate organic material are necessary. As these trees and shrubs come from a region of high rainfall, an occasional deep watering during summer droughts will prove beneficial. Generally fertilisers should be avoided, lime in particular being a cause of sudden demise in previously healthy trees.

PROPAGATION:
Either from seed, cuttings of semi-hardwood material or by removal of rooted suckers which spring up around mature trees.

USES:
A particularly colourful small tree for many garden situations, the brilliant flower display brightening the late spring scene. In the background of shrub borders or as a specimen, and to attract nectar-seeking birds, *E. coccineum* and its cultivars have found favour among gardeners in many parts of the world — particularly in Australia and New Zealand — but also in California and occasionally in parts of the United Kingdom.

PLATE 120: *Embothrium coccineum*.

Zoë C.

THE PROTEACEAE OF OTHER COUNTRIES

The previous sections detail and illustrate some of the better known and interesting Proteaceae that are under cultivation in various parts of the world. The following list acknowledges the other lesser known genera and their number of species currently described. New species are being discovered continually and have yet to be named or fully described.

Banksia; 1 — *B. dentata* also occurs naturally outside Australia in Papua New Guinea.
Beauprea; 10 in New Caledonia.
Beaupreopsis; 1 in New Caledonia.
Eucarpha; 2 in New Caledonia.
Finschia; 7 in New Guinea and the Pacific Islands.
Garniera; 1 in New Caledonia.
Grevillea; some in Pacific Islands, New Caledonia and South-East Asia.
Helicia; essentially a tropical species, apart from the 9 Australian members. Widely distributed throughout East and South-East Asia and Indomalaysia. Papua New Guinea has the greatest number of species at 47.
Heliciopsis; 7 in India to southern China and Western Malaysia.
Kermedecia; 10 in New Caledonia and Fiji.
Knightia; 1 in New Caledonia (1 other in New Zealand).
Macadamia; although the better known species are found in Australia, there are 4 others found in New Caledonia and South-East Asia.
Malagasia; 1 in Madagascar.
Oreocallis; a widely distributed genus with members from South America, but also with 1 in New Guinea and the South Moluccas.
Sleumerodendron; 1 in New Caledonia.
Stenocarpus; 18 in New Caledonia, 2 in Papua New Guinea (7 also in Australia).
Turrillia; New Guinea and Melanesia.
Virotia; 6 in New Caledonia.

GLOSSARY OF TERMS USED

acid: in reference to soil; with a pH value of less than 7.0. The lower the figure, the higher the acidity, e.g., pH 3 is strongly acid.
aeration: used in reference to soils comprising larger particles which allow air spaces between them. Soils with poor aeration are usually made up of fine particles with little air space and have a tendency to become waterlogged.
'alba': from the Latin *albus*, meaning white. Used to denote white-flowered variants.
alkaline: in reference to soil; with a pH value of more than 7.0. The higher the figure, the greater the alkalinity, e.g., pH 9 is strongly alkaline.
apex: the tip of an organ, such as a leaf.
'aurea': from the Latin *aurum*, meaning golden. Used to denote yellow- or gold-flowered variants.
awn: a fine, bearded or bristle-like appendage.
bract: a modified leaf. Usually surrounding a flower. Particularly used in reference to leucadendrons.
bypass: term used to describe new shoots arising from stem just below a flower and extending beyond the bloom while it is still opening. May be regarded as an undesirable feature for commercial purposes.
chlorosis: yellowing of foliage, usually due to a nutrient deficiency, but may also be due to a toxic condition.
clone: a group of individuals produced by vegetative asexual reproduction and therefore exactly alike.
cultivar: from *culti*vated *var*iety. One which has arisen, been cultivated, propagated, proved to be reliably persistent and named to comply with internationally accepted rules.
cultivation: term in broad use to describe conditions under which plants grow successfully in the home garden.
dioecious: having male and female flowers on separate plants.
entire: used in reference to leaves; with an even margin, without toothing or divisions.
erect: of upright habit.
foliage: leaves.
forms: term used loosely to cover different variants of a species.
fungus: a plant composed of thread-like filaments and having no chlorophyll so must secure foods from other plants or animals.
fungicide: chemical used to control fungous diseases.
genera: plural of genus.
genus: smallest natural group which contains related species and is intermediate between the family and species.
genetic: breeding origin.
glabrous: smooth, without hairs.
glaucous: blue-green in colouring and/or covered with a powdery bloom.
habitat: natural home of a plant or animal.
herbicide: chemical used in weed control.
humus: organic matter in an advanced state of decomposition.
hybrid: progeny resulting from the crossing between two species.
inflorescence: a cluster of flowers or a flowerhead.
insecticide: chemical used to control insect pests.
involucral: bracts or leaves surrounding flowers.
lanceolate: lance-shaped, narrow, tapering at both ends. Used in describing leaf shape.
lateral: occurring at the side of an organ.
lignotuber: swollen axis of a plant at the base of the stem possessing dormant buds capable of sprouting.
lignotuberous: persistent rootstock as in *Telopea speciosissima, Protea cynaroides* and other members of the Proteaceae.
luteum: from the Latin *luteus*, meaning yellow.
margin: the edge of a leaf.

mulch: loose material laid down to protect roots of plants. Also used to describe other soil coverings such as plastic.
neutral: used to denote pH 7. Neither acid nor alkaline.
nutrients: sources of food for plants.
oxidising: combining with oxygen, used in reference to blackening of protea leaves after harvesting of blooms.
pathogen: organism or substance causing disease.
pendent: hanging.
pendulous: of hanging habit, e.g., flowers of *Protea nana*.
perianth: the floral parts outside the stamens; calyx and corolla of the flowers.
pest: insect destructive to plants.
proteoid: term applied to fine rootlets of proteaceous plants.
pubescent: covered with short, soft or silky hairs.
reflex: bend or turn sharply backwards.
rootstock: a rhizome.
scale: 1. Term used to describe lower bracts on protea flowers.
2. Scale insects; hard, white or black flattened insects which fasten on to leaves.
semi-hardwood: growth which has begun to firm up, between spring and winter. Term applied to cutting material which is at the correct stage for propagating from, often in autumn.
species: singular or plural term to denote units within a genus that are distinguishable from each other.
stamen: pollen-producing part of a flower, comprising anther and filament.
style: slender part of the pistil in a flower, carrying the pollen-receiving stigma.
subspecies: between a species and a variety.
systemic: term used, in case of spray chemicals, to indicate those which are absorbed or translocate through the plant sap stream either through the roots or the foliage.
terminal: in flowerheads, relating to those borne at the tip or apex of the stem.
toxicity: term used for condition caused through excessive plant uptake of elements, e.g., phosphorus toxicity.
variant: form arising from a variation. A definable individual plant or group of, displaying differences from the normal species.
variety: plant which is subordinate to a species or subspecies and which has been validly and legitimately named and published in accordance with the rules of nomenclature.
vegetative: used in 'vegetative reproduction', meaning reproduction by means such as grafting, budding or cuttings.
veld: (from the Dutch *veld,* formerly *veldt*) open, unforested or lightly forested grass-country.
viable: in reference to seeds, those which are capable of remaining alive and germinating, or those which are fertile.
xylem: woody tissue within stem of a plant, comprised of vessels and fibres concerned with transportation of solutions from roots.

BIBLIOGRAPHY

Blombery, Alec & Maloney, Betty. 1981. *Proteaceae of the Sydney Region*. Angus & Robertson, Sydney.
Burke, Don. 1983. *Growing Grevilleas.* Kangaroo Press, Sydney.
Eliovsen, Sima. 1973. *Proteas for Pleasure.* MacMillan, Johannesburg.
George, Alex S. 1981. The Genus *Banksia*. *Nuytsia*, Vol. 3. No. 3. Western Australia.
________. 1984. *The Banksia Book*. Kangaroo Press, Sydney.
________. 1984. *An Introduction to the Proteaceae of Western Australia.* Kangaroo Press, Sydney.
Hocking, P. J. & Thomas, M. B. 1974. *Evolution of the Proteaceae and Cultural Implications.* Royal New Zealand Institute of Horticulture, Ann. 18-25.
Holliday, Ivan & Watton, Geoffrey, 1975. *A Field Guide to Banksias.* Rigby, Sydney.
Matthews, Lewis J. 1983. *South African Proteaceae in New Zealand.* Matthews Publishing, New Zealand.
________. 1993. *The Protea Growers Handbook.* David Bateman Ltd., New Zealand.
Nixon, Paul. 1987. *The Waratah.* Kangaroo Press, Sydney.
Rousseau, Frank. 1970. *The Proteaceae of South Africa.* Purnell, Cape Town.
Rourke, John P. 1972. Taxonomic Studies on *Leucospermum*. *Journal of South African Botany*, Vol. 8.
________. 1980. *The Proteas of Southern Africa.* Purnell, Cape Town.
Vogts, Marie M. 1982. *South Africa's Proteaceae; Know Them and Grow Them.* C. Struik (Pty) Ltd. Cape Town.
Williams, Ion J. M. 1972. A Revision of the Genus *Leucadendron* (Proteaceae). *Contributions from the Bolus Herbarium.* No. 3:105, Cape Town.
Wrigley, John W. & Fagg, Murray. 1989. *Banksias, Waratahs and Grevilleas and all other plants in the Australian Proteaceae Family.* Collins Publishers, Sydney, N.S.W.

INDEX